Mehdi Salehi

Bayesian-Based Predictive Analytics for Manufacturing Performance Metrics in the Era of Industry 4.0

Reihe Informationsmanagement im Engineering Karlsruhe
Band 2 – 2019

Herausgeber
Karlsruher Institut für Technologie
Institut für Informationsmanagement im Ingenieurwesen (IMI)
o. Prof. Dr. Dr.-Ing. Dr. h.c. Jivka Ovtcharova

Eine Übersicht aller bisher in dieser Schriftenreihe erschienenen Bände finden Sie am Ende des Buchs.

Bayesian-Based Predictive Analytics for Manufacturing Performance Metrics in the Era of Industry 4.0

by
Mehdi Salehi

KIT Scientific Publishing

Dissertation, Karlsruher Institut für Technologie
KIT-Fakultät für Maschinenbau

Tag der mündlichen Prüfung: 7. Dezember 2018
Referenten: Prof. Dr. Dr.-Ing. Dr. h. c. Jivka Ovtcharova
Prof. Dr.-Ing. Rüdiger Haas

Impressum

KIT Scientific Publishing

Karlsruher Institut für Technologie (KIT)
KIT Scientific Publishing
Straße am Forum 2
D-76131 Karlsruhe

KIT Scientific Publishing is a registered trademark
of Karlsruhe Institute of Technology.
Reprint using the book cover is not allowed.

www.ksp.kit.edu

Print on Demand 2019 – Gedruckt auf FSC-zertifiziertem Papier

ISSN 1860-5990
ISBN 978-3-7315-0908-0
DOI 10.5445/KSP/1000091806

“Human beings are not born once and for all on the day their mothers give birth to them, but ... life obliges them over and over again to give birth to themselves.” – Gabriel García Márquez

Dedicated to my parents

Abstract

Industry 4.0 refers to the incorporation of a variety of technologies and agents in order to improve the efficiency and reliability of manufacturing systems. The Internet of Things, cyber-physical systems, data analytics, machine learning, and robotics are some of the components of the fourth industrial revolution. Recent developments in the era of Industry 4.0 are intelligent manufacturing systems including intelligent machine tools and related sensors, as well as design and manufacturing software packages. In this context, accurate prediction of performance and reliability of the machine tool components such as cutting tools is necessary to make intelligent decisions for producing of high-quality products. Predictive analytics as a branch of data analytics is widely used for modeling and prediction of machine tools and cutting tools performance metrics. The performance metrics such as cutting force and tool life are important factors, which influence the productivity of manufacturing processes. However, actual predictive analytics techniques (*e.g.*, regression and machine learning techniques) are mainly deterministic, so they do not take into account the inherent uncertainties and variabilities in the manufacturing process.

The research in this dissertation proposes Bayesian-based predictive analytics for the modeling and prediction of the performance metrics in machining processes including cutting force, tool life and reliability, and tool wear growth. Bayesian inference is a probabilistic method, which can model and minimize manufacturing process uncertainties. Using Bayesian inference initial belief or expert opinion can be integrated into the experimental data to predict manufacturing variables. To illustrate the applicability of Bayesian-based predictive analytics to the performance metrics, Bayesian Markov Chain Monte Carlo approach is applied to four design and manufacturing process applications. In the first application, probabilistic prediction of cutting force is performed, and the effect of cutting tools geometries are studied in an orthogonal turning process. The second application investigates prediction of cutting and ploughing forces using the probabilistic method in the turning process. The third application discusses probabilistic prediction of cutting tools life and reliability for different tool geometries and cutting data in the milling process. In the fourth application, the Bayesian inference is applied to predict tool wear growth using various tool rake angles in the milling process.

The objective of this research is to propose a probabilistic modeling approach, which can quantify manufacturing process uncertainties, integrates machining models with experimental data to infer the performance metrics, and finally incorporates historical data to current and future analysis in a sequential manner.

The output of this research provides the applicability of Bayesian methodology to the area of product design and manufacturing process. In this regard, the probabilistic approach can reduce the cost of the expensive and hazardous experiments by incorporating past and current information into the future analysis.

Kurzfassung

Industrie 4.0 bezieht sich auf die Integration einer Vielzahl von Technologien, um die Effizienz und Zuverlässigkeit eines Fertigungssystems zu verbessern. Das Internet der Dinge, cyber-physischeSysteme, Datenanalyse, Maschinelles Lernen und Robotik sind einige der Komponenten der vierten industriellen Revolution. Aktuelle Entwicklungen in diesem Kontext betreffen intelligente Fertigungssysteme, einschließlich intelligenter Werkzeugmaschinen und entsprechender Sensorik, sowie Design- und Fertigungssoftwarepakete. In diesem Zusammenhang ist eine exakte Vorhersage der Leistung und Zuverlässigkeit der Werkzeugmaschinenkomponenten, z. B. von Schneidwerkzeugen, erforderlich, um eine gleichbleibend hohe Fertigungsqualität sicherzustellen. Ein Schwerpunkt innerhalb der Datenanalyse ist die prädiktive Analyse. Diese wird häufig zur Modellierung und Vorhersage von Leistungs- und Zuverlässigkeitsindikatoren von Werkzeugmaschinen und der zugehörigen Fertigungsprozesse angewandt. Tatsächliche prädiktive Analysetechniken (z. B. Regressions- und Maschinelle Lerntechniken) sind jedoch hauptsächlich deterministisch, sodass sie die inhärenten Unsicherheiten und Variabilitäten des Fertigungsprozesses nicht berücksichtigen.

In dieser Arbeit wird daher eine Bayesian-basierte prädiktive Analyse zur Modellierung und Vorhersage von Leistungs-, Zuverlässigkeits-, und Verschleißmerkmalen von Zerspanungsprozessen als Forschungsansatz vorgeschlagen. Die Bayes'sche Inferenz ist eine probabilistische Methode mit den Unsicherheiten des Fertigungsprozesses modelliert und anschließend minimiert werden können. Mit Hilfe dieser Methode können experimentelle Ergebnisse in Verbindung mit Expertenwissen zur Prognose verschiedene Fertigungsgrößen genutzt werden.

Um die Anwendbarkeit der Bayes'schen prädiktiven Analyse für die Leistungs- und Zuverlässigkeitsindikatoren des Fertigungsprozesses zu demonstrieren, wird ein Bayes'sche Markov-Chain-Monte-Carlo-Ansatz auf vier Design- und Fertigungsanwendungen genutzt.

Zunächst wird eine probabilistische Vorhersage der Schnittkraft in einem orthogonalen Drehprozess durchgeführt und der Effekt der Schneidwerkzeuggeometrien untersucht. Die zweite Anwendung untersucht die Vorhersage von Schnitt- und Reibungskräften unter Verwendung eines erweiterten Modellierungsansatzes im Drehprozess. Die dritte Anwendung behandelt die probabilistische Vorhersage der Lebensdauer und Zuverlässigkeit von Schneidwerkzeugen für verschiedene Werkzeuggeometrien und Schnittdaten in einem Fräsprozess. Abschließend wird der Ansatz angewendet, um das Werkzeugverschleißwachstum unter Verwendung verschiedener Werkzeugschneidwinkel beim Fräsprozess vorherzusagen.

Das Ziel dieser Forschung ist es, einen probabilistischen Modellierungsansatz zu entwickeln, der Unsicherheiten in Fertigungsprozessen quantifiziert und Bearbeitungsmodelle mit experimentellen Daten integriert. Damit können Leistung, Lebensdauer und Zuverlässigkeit des Fertigungsprozesses abgeleitet werden. Zur Steigerung des Modellierungsgrad werden zusätzlich historische Daten in laufende und zukünftige Analysen integriert.

Das Ergebnis dieser Forschung liefert die Anwendbarkeit der Bayes'schen Methodik für den Bereich des Produktdesigns und des Herstellungsprozesses. In dieser Hinsicht kann der probabilistische Ansatz die Kosten und Anzahl von Versuchen reduzieren, indem aktuelle und bereits bekannte Informationen in zukünftige Analysen einbezogen werden können.

Acknowledgment

Firstly, I would like to express my sincere gratitude to my advisor Prof. Dr. Dr.-Ing. Dr. h. c. Jivka Ovtcharova for her continuous support of my doctoral study and related research, for her patience, motivation, and immense knowledge. She guided me throughout my research, and in writing of this thesis

My sincere thanks also go to my second advisor, Prof. Dr.-Ing. Rüdiger Haas for his generous support and guidance. I have grown both professionally and personally under his guidance. I also thank my colleagues at IMP institute of the Hochschule Karlsruhe Technik und Wirtschaft and my industry partner Zermet Zerspanung GmbH for their technical support and continuous encouragement.

I am also grateful to my co-advisor Prof. Tony Schmitz, and my colleagues at the University of North Carolina at Charlotte, Ryan Copenhaver, Chris Kossack and Andrew Honeycutt. I am very thankful to Dr. Schmitz for sharing his expertise, providing me with an opportunity to research Bayesian statistics with his research group, and for his valuable guidance and encouragement extended throughout my thesis.

I take this opportunity to express gratitude to Karlsruhe House of Young Scientists (KHYS) at KIT for their financial support during my research stay at the University of North Carolina at Charlotte. I am also thankful to Stifterverband and Doktorandenkonvent of the Hochschule Karlsruhe Technik und Wirtschaft for providing financial aid for the experiments and equipment and documentation.

Last but not the least, I would like to thank my family: my parents, my brothers, and sisters and my friends, especially, Mehdi Omidvar, Gabriel Wald and Niclas Zeller, for supporting me technically and spiritually throughout writing this thesis and my life in general.

Declaration

I am aware of and understand the policy of Karlsruhe Institute of Technology (KIT) on plagiarism, and I certify that this thesis is my work, except where indicated by referencing and the work presented in it, has not been submitted in support of another degree or qualification from this or any other university or institute of learning.

The data presented in this thesis was obtained in the experiments carried out by myself at the Hochschule Karlsruhe Technik und Wirtschaft, and University of North Carolina at Charlotte, USA. I played a significant role in the preparation and execution of the experiment, and the data analysis and interpretation are entirely my own work.

Hiermit versichere ich, dass ich die Richtlinien des Karlsruher Instituts für Technologie (KIT) hinsichtlich Plagiate kenne und bestätige, dass ich die vorliegende Arbeit selbstständig verfasst habe, sofern Inhalte nicht durch Verweise und darin präsentierten Arbeiten zur Unterstützung ausgewiesen sind. Diese wurden nicht zum Erreichen eines anderen Abschlusses/ einer anderen Qualifikation von dieser oder einer anderen Universität oder einer anderen Hochschule eingereicht.

Die in dieser Arbeit präsentierten Daten wurden in Studien von mir an der Hochschule Karlsruhe-Technik und Wirtschaft und an der University of North Carolina in Charlotte, USA, gewonnen. Die Vorbereitung und Durchführung der Studien wurden unter meiner Anweisung durchgeführt. Die Datenanalyse und -interpretation erfolgte ausschließlich durch eigene Arbeit.

Karlsruhe, den 20.09.2018
Mehdi Salehi

List of Figures

List of Tables

Contents

List of Symbols

Symbol	*Description*	*Units*
θ	Model parameter	
$p(\theta)$	Probability density function of parameter θ	
$p(\theta \vert y)$	Posterior of parameter θ given data y	
$p(y \vert \theta)$	Likelihood of the data y given parameter θ	
θ_{ML}	Maximum Likelihood of parameter θ	
θ_{MAP}	Maximum a posteriori	
σ	Standard deviation	
F_t	Tangential force	*N*
F_f	Feed force	*N*
K_t	Merchant tangential force coefficient	*MPa*
K_f	Merchant Feed force coefficient	*MPa*
b	Width of cut	*mm*
h	Uncut chip thickness	*mm*
τ_s	Shear stress	*MPa*
α_r	Rake angle	*deg*
F_s	Shear force	*N*
β_a	Average friction angle	*deg*
ϕ_c	Shear plane angle	*deg*
K_{tt}	Kienzle tangential force coefficient	*MPa*
K_{ff}	Kienzle Feed force coefficient	*MPa*
c_t	Kienzle tangential force exponent	
c_f	Kienzle feed force exponent	
R_a	Arithmetical man roughness	*μm*
R_z	Ten points mean roughness	*μm*
f	Feed	*mm/rev*
V_c	Cutting speed	*m/min*
A_s	Shear area	mm^2
r_c	Merchant chip ratio	
h_c	Merchant cut chip thickness	*mm*
K_{te}	Kienzle ploughing force coefficient	*MPa*
δ	Ploughing layer	*mm*
F_{tp}	Ploughing force	*N*
α_s	Locating angle	*deg*

h_m	Extended Kienzle cut chip thickness	*mm*
C	Taylor constant	*m/min*
n	Taylor exponent	
T	Tool life	
$V_{B,avg}$	Average tool flank wear	*mm*
$V_{B,max}$	Maximum tool flank wear	*mm*
$V_{B,N}$	Tool notch wear	*mm*
$V_{B,c}$	Tool corner wear	*mm*

List of Acronyms

ACR	*Acronyms*
ACO	Ant colony optimization
AI	Artificial Intelligence
ANN	Artificial Neural Network
CAD	Computer-Aided Design
CAE	Computer-Aided Engineering
CAM	Computer Aided Manufacturing
CAPP	Computer Aided Process Planning
CDF	Cumulative Distribution Function
CNC	Computerized Numerical Control
CPPS	Cyber-Physical Production Systems
CPS	Cyber-Physical Systems
DA	Data Analytics
FEA	Finite Element Analysis
FEM	Finite Element Method
FDM	Finite Difference Method
FL	Fuzzy Logic
FORM	First Order Reliability Method
GA	Genetic Algorithm
GDP	Growth Domestic Product
IoT	Internet of Things
IIoT	Industrial Internet of Things
LSF	Least Squares Curve Fitting
MAP	Maximum a Posteriori
MCMC	Markov Chain Monte Carlo
ML	Maximum Likelihood
MH	Metropolis-Hastings Algorithm
MRO	Maintenance, Repair and Operations Management
NN	Neural Network
PDF	Probability Density Function
PDM	Product Data Management
PLM	Product Lifecycle Management
PSO	Particle Swarm Optimization
RMS	Reconfigurable Manufacturing System
RV	Random Variable

SA	Simulated Annealing
SO	Smart Objects
SD	Smart Data
SP	Smart Product
SRF	Stochastic Response Surface
VA	Virtual Assembly
VMach	Virtual Machining
VMT	Virtual Machine Tool
VTo	Virtual Tooling
VP	Virtual Prototype

Part I: Introduction to the Research

1 Industry 4.0 and Predictive Analytics

1.1 Industry 4.0

The industrial revolution began with the transition from hand production to machine production in Great Britain. According to Figure 1.1, Industry 1.0 corresponds to the first step in industrialization, in which steam-powered machines and mechanisms replaced human and animal powered machinery in production facilities. The first industrial revolution spanned from the end of the 18th century to the beginning of the 19th century. Industry 2.0 was mass production starting around 1870 but is best known through the assembly lines of Henry Ford at 1913. Industry 3.0 corresponds to the introduction of computers and automation in manufacturing from 1950 onward.

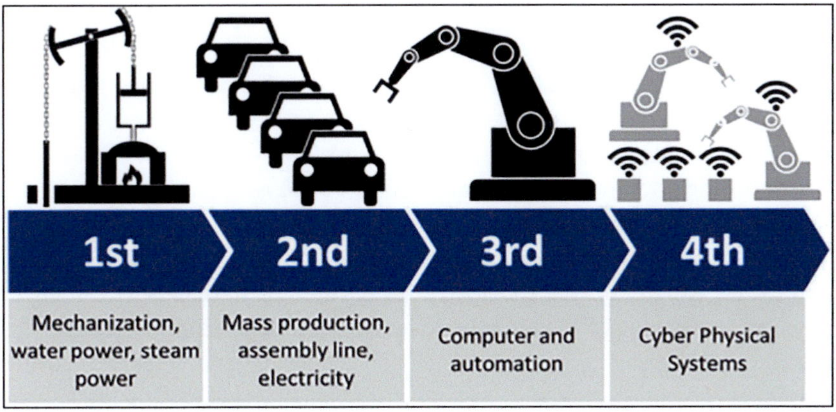

Figure 1.1: Industrial revolutions and prospects [1]

Industry 4.0 refers to the incorporation of a multiplicity of technologies that evolved from a computer-controlled automated facility into a system that gathers and analyzes data from the floor to make intelligent decisions in an automated manner. The Internet of Things (IoT), big data, cyber-physical systems

(CPS), machine learning (ML), and robotics are some of the components that are associated with this revolution, which changes today's production technology to cyber-physical production systems (CPPS). Industry 4.0 and its components can improve efficiency, agility, and reactivity to the rapidly changing market demand [1–3].

The term "Industry 4.0" was employed for the first time in 2011 at the Hannover Fair in Germany. It encompasses some modern automation, data exchange, and manufacturing technologies and has been defined as follows [4]: "the collective term for technologies and concepts of value chain organizations which draw together CPS, the IoT and the Internet of Services."

The emergence of low-cost sensors and actuators, and communication through the internet, enables real-time connections between machines, products, end users, which is defined as the IoT technology. Today, IoT empowers manufacturing systems to collect more and more data from machines and processes. In this scenario, the rapid development of data analytics approaches improves the diagnostics and prognostics of production systems, significantly. Data analysis methods provide better knowledge of the manufacturing system in online and offline status. Additionally, cloud-based computation methods are developing rapidly. This allows manufacturing systems to develop and deploy more efficient predictive analytics, prognostics and diagnostics techniques, such as online monitoring [5], indirect estimation of manufacturing technological indicators (*e.g.,* cutting forces and vibrations) [6,7] and remaining useful life predictions [8]. Another component of Industry 4.0, which has significant potential for today's manufacturing systems is CPS. A CPS has two elements, physical and cyber, which are interconnected. While the real system operates in the physical world, the digital system operates in the cloud platform, simulating the machine health monitoring, and continuously recording and tracking machine conditions and product quality, among other valuable information processing. Consequently, data-driven models can be developed to integrate the information from physical systems into the digital models [9].

The IoT, data analytics, cloud computing, and CPS have developed a new paradigm in manufacturing called personalized production. Accordingly, the end users of the products can be involved with the customized design process from the early stages until the services life and disposal [10]. This can be achieved with the development of a new generation of smart factories, which are characterized by flexible production systems and reconfigurable manufacturing system (RMS) that allow the successful personalized production of products even in small batches. Using RMS allows building a *live* factory where its structure changes cost-effectively in response to markets and end-user demands for rapid responsiveness to unexpected products changes [9].

Industry 4.0 refers to a new approach for organization and control of value-adding systems; see Figure 1.2. The principal objective of Industry 4.0 is to provide individual customer demands at the cost of mass production. In this context, all areas including order management, research and development, production, and supply chain are influenced. The sub-structure for such a contribution can be developed by the digitalization of production using cyber-physical production systems (CPPS). In this way, all of the resources, including workers, products, and machines have to be integrated as intelligent, self-organized, interconnected, real-time and autonomously optimized instances [3].

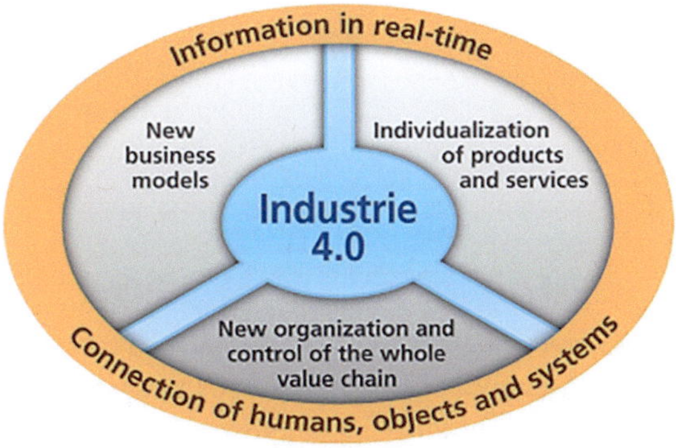

Figure 1.2: Industrie 4.0 [11] (qtd. in [3])

As illustrated in Figure 1.3, there are a variety of technologies and components that are integral to Industry 4.0. Four of them (CPS, product lifecycle management (PLM), virtual manufacturing system, and data analytics) are described here, all of which are relevant to the current research.

1.1.1 Cyber-Physical Systems

Cyber-physical systems are an inevitable outcome of the information revolution. Embedded computing systems, internet communication, and digital technology are integral components of modern industries. The National Science Foundation (NSF) of the United States has been performing fundamental research in CPS since 2010. According to NSF, CPS is defined as: "engineered systems that are built from, and depend upon, the seamless integration of computational algorithms and physical components." CPS will operate and control innovation and competition in various sectors, *e.g.*, agriculture, transportation, construction automation, energy, medical service, and manufacturing [4].

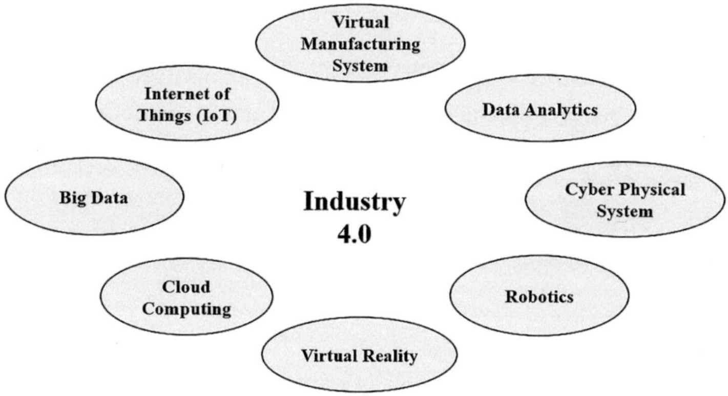

Figure 1.3: Components of Industry 4.0

The term *cyber-physical* encompasses the essential characteristic of the following concept: the integration of embedded systems to observe and control physical processes in the framework of digital network technology. Different types of sensors and actuators compound a networked structure of control systems for near-real-time data processing. Using this structure, autonomous decisions can be made, so that the system can adapt itself to new surrounding conditions to a certain level. This type of system is already utilized in various applications such as cell phones, aerospace, and energy power plants among others [4,12].

Development of the CPS in future production systems paves the way to build a *smart factory*. In a smart factory, the production facilities and logistics systems largely organize themselves without human intervention. Part of this future scenario continues to be the communication between the product (*e.g.,* workpiece) and the production plant, in which the product itself brings its manufacturing information in machine-readable form, on radio-frequency identification (RFID) chip. These data are used to control the path of the product through the production facilities and the manufacturing process steps. As an example, the use of RFID-based cyber-physical systems in a clothing industry led to increase of production efficiency by 25 %, where the RFID chips could recognize the bottleneck operation and measure the working time of each operator in the manufacturing process [4].

1.1.2 Product Lifecycle Management

Industry 4.0 involves the development and production of products for consumers using the IoT, machines and production plants. This can be actually done by incorporating the industrial value chain within the entire lifecycle of the product using digital technology. In this regard, Product Lifecycle Management (PLM) can be the backbone of this digitalization. PLM is a principal source for all information about a product, from the initial concept (through engineering design) and production to service and disposal [13]. PLM incorporates people, information, production processes, and business systems and provides information about the performance of the product to companies and their extended enterprise [14]. There are several PLM models in the industry, but most of them are similar. In general, a PLM model of a physical product may have three phases, as follows [15]:

1. **Design and development:** In this phase, first, the product design specifications and requirements are studied. Next, the conceptual design is conducted considering the functions and features definition for the products. The initial analysis is performed using tridimensional models and tolerance analysis. The detailed design is performed by material and functionality analysis of the products. The software packages for the feature and functionality design are called computer-aided tools (CAx) for engineering applications. Ultimately, manufacturing process planning is done in micro and macro levels generally using computer-aided production planning (CAPP) software programs.

2. **Manufacturing:** Once the design and development of the products are finalized, the manufacturing phase is defined. This includes defining production methods, techniques and processes to produce the parts. The software programs for this phase are computer aided manufacturing (CAM) packages. Eventually, the assembly process is performed, and the products are prepared for the storage.
3. **Service life, operation, and recycling:** The final phase of the lifecycle deals with the distribution, operation, user-product interaction, and recycling and reuse of the products. This can include providing customers with the support and information required for maintenance, and product functions and feature optimization using tools such as maintenance, repair and operations management (MRO) software. There is end-of-life to every product. Whether it be disposal or recycling of material objects or information, this needs to be carefully considered since it may not be free from consequences.

1.1.3 Virtual Manufacturing System

The conceptual design has been exercised since the 1960s with the introduction computer-aided design (CAD) and computer aided manufacturing (CAM). Engineering applications analysis has also conducted using computer-aided engineering (CAE) tools, *e.g.*, finite element analysis (FEA) [16]. Since the 1990s, a paradigm shift in manufacturing from *real* to *virtual* manufacturing has happened, which resulted in significant interests for researchers. The computers allow simulating the physical manufacturing systems in virtual environments. The virtual design and manufacturing packages can emulate the behavior of manufacturing systems on a computer before physical production, and therefore it reduces the number of experiments on the shop floor. Virtual systems, including virtual machine tools (VMT), virtual machining (VMach), virtual assembly (VA), virtual tooling (VTo) and virtual prototype (VP) have been developed to support virtual manufacturing systems. The virtual systems enable the manufacturers to consume less wasted materials and production interruptions in the plants. The systems can avoid or prevent hazards regarding the manufacturing processes and machines. Additionally, product data management (PDM) and product lifecycle management (PLM) can be carried out more systematically [17].

Several virtual machining research works have been conducted to model, simulate, visualize and predict the process performance metrics (such as cutting forces, tool wear) early in the process planning stage and increase the quality of the manufactured parts, such as surface quality and tool vibration [16]. For example, Ehmann *et al.* [18–20] developed a virtual machining system to (1) approximate the size effect using a method to calculate cutting-condition-independent coefficient and prediction of cutting forces over a wide range of cutting conditions, (2) to predict the three-dimensional machined surface errors generated during the peripheral end milling process and (3) predict cutting forces in the transient cut. Virtual optimization models have also been developed to be able to select the machining parameters such as speeds, feed rates, and depth of cut in a proper manner. This can improve process efficiency and quality of the finished parts.

There are numerous VMT models and application developed to construct machine tools and provide a *virtual prototyping* environment [16]. The applications and methods aim to reduce the cost and time for development and improvements of the physical prototypes. For example, Altintas *et al.* [21] proposed a virtual prototype model for machine tools to modify the virtual design, iteratively. In this context, the virtual porotype approach enables the engineers to simulate the kinematic, static and dynamic of the machine. The model also allows the designer to avoid optimization of the physical machine prototype, based on trial and error, and experiences.

1.1.4 Data Analytics

Nowadays, manufacturing companies are growingly faced with the influences of dynamic workflows and the development of continually increasing process complexity. In order to do this, the companies have to adjust their production machines and processes according to their customers' requirements in the supply chain. In this context, production control is especially suffering from issues of low prediction quality in production planning. The reasons for the inaccurate prediction quality include the uncertainties of future clients' demands, the unpredictability of the machine's downtimes, and unreliability of the products sub-suppliers, among others. Currently, the solutions to solve the inaccurate prediction are provided by a variety of data analytics and IT-systems, which, mostly, are not efficient due to noisy or biased data [4].

In order to come up with a solution to the mentioned problems, one can propose to implement various approaches from the area of data analytics. Due to the increase of demand for digitalization of the industries, data processing becomes more significant. In this context, essential factors for an increase of productivity are automation of the production lines and application of the intelligent and efficient data analytics tools. This requires implementation of the technologies from the area of data acquisition, data processing, data assessment, and data exchange.

Data acquisition alone is incomplete and not profitable for industries. The process can be interesting for companies and users when historical data are used to predict manufacturing future measures and conditions. This can be done using self-learning algorithms. In this regard, smart data (SD) can play an essential role in the transparency of the manufacturing process and product supply chain, and it can help the companies in making informed decisions [4].

The data analytics has been traditionally practiced much more in the area of customer service than in the production industry. Hence, manufacturing industries can benefit from the already developed data analytics methods as inputs to their applications for innovative solutions. In this context, a derived method of data analytics in production is described in Figure 1.4 [4].

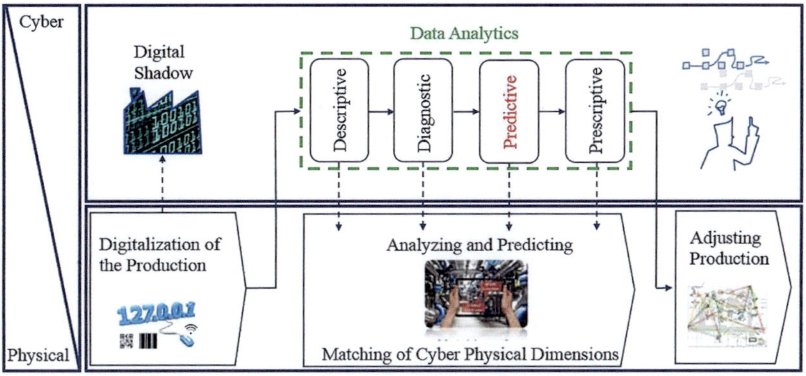

Figure 1.4: Data analytics as the enabler for CPPS [22] (qtd. in [4])

The above figure demonstrates data processing stages using diagrams within a cyber-physical production system (CCPS). A vital requirement for CCPS is the digitalization of the manufacturing process, which can be performed using a digital shadow of the process. The Digital shadow resembles an abstract image of the manufacturing system. This is a necessary step to generate information for data analytics. The objective of this phase is to define essential data sources including the measurable and understandable data for information technology and to incorporate all sources of information.

In the second phase, the data analytics approach, which is derived from business and customer service is implemented. In order to utilize the approach efficiently, it has to be reshaped according to the production process requirements. Four successive steps of the data analytics are described as follows [4]:

1. **Descriptive analytics:** The first stage of data analytics concentrates on increasing transparency of the production using the data generation. In order to control and optimize the production process, it is necessary to acquire a digital shadow. This can be only done using fully equipped production with sensor technology, such as a barcode, RFID, cameras.
2. **Diagnostic analytics:** Diagnostic analytics is a type of advanced analytics which investigates data to answer the question, why did it happen? Moreover, is characterized by some techniques such as drill-down, data

discovery, and data mining. Diagnostic analytics considers a more in-depth exploration of data to infer the causes of events and behaviors [23].

3. **Predictive analytics:** In the third phase, the objective is to make predictions using the previously identified patterns. Predictions exit in everyday life such as weather forecasts or the rear-sensors of a car. In production, forecasts are relevant, when, for example, a bottleneck situation is foreseen or when a machine breakdown would not occur yet. Predictive analytics provides excellent potential for production planning to achieve a smooth production process.
4. **Prescriptive analytics:** The last step of the data analytics approach is prescriptive analytics. It is used for making decision support for many organizational problems and therefore enables quantified and reasonable decisions for current and future manufacturing applications.

The last stage of the CPPS is the adjusting production. In this phase, the output information of the data analytics can be visualized in the form of new assistance systems (*e.g.*, tablets or virtual reality support systems). This can provide information to the employees to adjust production and implement the derived suggestions by the system based on results of the data analytics. For example, using the recommendations for the production planner, bottlenecks can be avoided, or orders can be rearranged due to the lack of parts. It is important to note that the decision proposals for the production planner, made by the CPPS support, should be made for the entire production system since the production units and the processes are interconnected in a CPPS system [4]. Although all the mentioned stages of data analytics are important in coming up with a concrete solution for the improvement of the decision-making process, the focus of this thesis is on predictive analytics in the manufacturing process.

1.1.5 Predictive Analytics

Predictive analytics uses statistics approaches to identify meaningful patterns, which can predict the likelihood of future outcomes based on historical data. Predictive analytics utilizes techniques such as data mining, statistics, probability theory, machine learning, and artificial intelligence to detect trends and patterns in the measured data. The patterns found in the historical data can be used to identify the risk and probability of occurrence of an event in the future.

Predictive analytics enables the companies and organizations to predict behaviors of a system or a process based on the data (quantitative technique) and not on the assumption and hypothesis (qualitative analysis). Predictive analytics have various applications and usage in marketing, financial services, production. Some of the applications are as following [24]:

1. **Optimizing product quality:** Predictive analytics enables manufacturers to improve products quality. In this context, by accurately detecting patterns from the information related to the production problems and rapidly discovering their causes, manufacturers can forecast and decrease the number of defects earlier than they influence their customer's satisfaction, safety, and confidence.
2. **Enhancing warranty planning:** Using predictive analytics, designers and manufacturers can identify failure times of machines parts or products earlier than their occurrence. In doing so, smarter decision in defining service and repair policies can be made to ensure profitability in warranty practices.
3. **Improving demand planning and inventory management:** Predictive analytics provides designers and manufacturers broader vision about sales trends and customers' demands. Hence, production output can be organized to avoid stock shortages or overages due to the miscalculation of the market demands, and smarter decisions can be made regarding the procurement of components and raw materials. In such a way, more economical business models, such as just-in-time or lean manufacturing strategies, can be implemented.
4. **Increasing maintenance:** Manufacturers implement inspection schedules for their machinery to keep production output constant. Predictive analytics can assist them to understand their equipment failure times and reasons better. This provides them with more proactive approaches for the maintenance of the machines, so that they ensure the machinery performance, and avoid downtime and production delays, in addition to costly and unnecessary maintenance.

Predictive analytics for manufacturing systems enables users transparency in operations. The basis of predictive manufacturing is intelligent software, which is used to control the functionalities of predictive modeling. Using predictive modeling methods, manufacturers can obtain the opportunity to proactively carry out appropriate solutions to avoid losing efficiency in manufacturing operations. Predicting the performance of production equipment and the estimation of their failure time can mitigate the effects of these uncertainties. Predictive analytics provides manufacturing companies the possibility to predict their machines performance before breakdowns happen using recognized patterns which are achieved from the past machine failures. With these patterns and data, a manufacturer can perform the necessary maintenance schedules on machinery during slow periods without interrupting production operations [25].

As a predictive analytics tool, artificial neural networks (ANNs) are widely used for intelligent data processing. The ANNs are already utilized to predict changes in complex systems such as weather conditions forecast or the prediction of time series for economic growth problems. Information technology attempts to demonstrate the human nervous system with the help of neural networks. The learning ability is made by the independent activation of connections in the network, changes of weightings, and adding and removing of neurons [4,26].

For predictive systems, access to high-resolution data from production is an essential requirement. Using this data, the ANNs learn the coherence of historical events and can predict the impact on future decisions. The ANNs can be implemented in various fields of decision making for the production process control. They can also help the decision-makers for diagnostic analysis in the field of pattern recognition, control theory and robotics [27].

1.1.5.1 Predictive Analytics Techniques

Predictive analytics approaches and techniques can mainly be classified into regression and machine learning techniques, which are described in the following sections.

Regression Techniques

Regression techniques are the pillar of predictive analytics. The modeling technique is a useful approach for different types of inferential tasks such as prediction, parameter estimation, and data description. The objective is to establish a mathematical equation as a model to represent the interaction among variables of the model. For instance, a linear regression model investigates the relationship between the dependent variable and a set of independent variables. This relationship is defined as an equation that predicts the dependent variable as a linear or nonlinear function of the independent parameters. These parameters are determined to optimize the regression quality (*i.e.,* the goodness of fit). Regression is commonly used for model fitting technique, which concentrates on minimizing the residuals [28].

Machine Learning

Machine learning (ML) is a branch of artificial intelligence that systematically implements algorithms to incorporates the underlying relationships within data and information. In 1959, Arthur Samuel described machine learning as "field of study that gives computers the ability to learn without being explicitly programmed" [29]. ML employs several advanced statistical techniques for regression and classification. In this regard, there are two significant types of training methods used by ML including supervised and unsupervised. The supervised learning techniques extract relationships between independent variables and designated dependent variables. It utilizes a training dataset to develop a prediction model using input data and output values. The model can then be used to make predictions of the output values for a new dataset. Unsupervised learning techniques group datasets without a prespecified dependent variable. This technique often involves learning structured patterns in the data by rejecting pure unstructured noise. The unsupervised learning can be further grouped to clustering and association algorithms. The applications of machine learning are increasingly developed in a variety of fields including medical diagnostics, engineering, multimedia analysis, and the stock market, among others. There are different approaches to machine learning such as decision tree, neural networks, deep learning, and support vector machines (SVMs) [30,31].

1.1.5.2 Challenges of Predictive Modeling

The predictive data analytics in manufacturing is becoming increasingly important and having more significant impact on a wide range of industries. However, there are many uncertainties and issues associated with the data analytics, which are common to all industries. In this regard, some of the challenges are listed as following [32]:

1. Data quality is labor-intensive. Having a massive amount of data does not necessarily result in a better output quality. Typically, most data scientists spend 75%-80% of their time cleaning up data. Analyzing poor quality data causes misleading the information.
2. Data Reliability relies on collection methods and the definition of measurements process. Data reliability depends on how the raw data was gathered. The robustness of analysis is related to data quality and reliability.
3. Variability in the manufacturing process leads to different data even though the process parameters are identical. The variability originates from inherent uncertainty in the manufacturing system such as material-to-material differences and changes in physical behaviors of the production machines. Predictive analytics tools should be able to quantify the uncertainty of the manufacturing processes.
4. As stated by Box [33]: "essentially, all models are wrong, but some are useful". A model's potential for inaccuracy is an important consideration in predictive modeling. Even a correctly specified model may not always be accurate. An error term represents the portion of the model that is unexplained. Model uncertainty is the uncertainty related to imperfect knowledge or idealizations of the mathematical models. It is typically difficult to characterize the modeling errors in an efficient and statistically consistent manner using classical statistical techniques (*e.g.*, regression analysis).
5. Predictive models are often based on frequentist (classical) approaches, which are based on testing the null hypothesis. In other words, the null hypothesis implies that no information is available. This hypothesis could be an inappropriate starting point for data analysis because based on previous research, it is almost always expected that some information is available.

1.1.6 Bayesian Method for Predictive Analytics

Bayesian method enables users to integrate background knowledge into their analyses instead of testing the same null hypothesis repeatedly and ignoring the learnings from previous experiments. The Bayesian method is a knowledge-based approach and differs from data-driven approaches which are based on the null hypothesis. Bayesian approach to predictive modeling incorporates prior knowledge or initial belief (*i.e.*, past information or expert's knowledge) about a parameter into the data from current experiments to form a posterior knowledge of the parameter. The posterior knowledge (the updated parameter) can be used as a prior belief for the next set of analysis, sequentially. This can minimize the number of experiments in case the tests are time-consuming or costly.

While the majority of the predictive modeling techniques are deterministic, Bayesian approach can model and quantify inherent uncertainties of the manufacturing processes. The approach is used to minimize uncertainties. This is achieved by training model parameters using the results of experiments in the context of the likelihood function. Using the Bayesian method, the degree of confidence or belief about a model parameter is maximized after training the parameter with the results of new observation and experiments.

Another advantage of the Bayesian approach for predictive modeling is that it is highly flexible. Using the Bayesian method, it is straightforward to fit rational models to complex datasets with measurement error, censored or missing observations [34].

The Bayesian method has also been used for the probabilistic design of industrial components, where the prediction of the probability of failure in addition to safety and risk assessment are utmost important. The approach can be applied to various industries. In this regard, some of the applications are probabilistic risk assessment in nuclear plant [35], the probabilistic design of aircraft turbine disk at high pressure [36], and probabilistic design of wind turbines [37] among others. The probabilistic design is characterized by prediction of reliability and failure probability of the designed components using statistical distributions. This differs from deterministic design, which is based on using safety factors for industrial components design.

1.2 Case Study: Bayesian-Based Predictive Analytics for Machining Metrics

Cutting tools are the most flexible components in a manufacturing system. The quality of the different machining processes is affected by cutting tools performance and reliability. An important machining and cutting tools performance metric is the cutting force. There are some machining technological indicators (*e.g.*, machining stability and vibration, tool wear growth, and machining power consumption), which have been directly or indirectly attributed to the cutting force [38]. Online monitoring and prediction of the cutting force can avoid damages to the machine tool components such as cutting tools and spindle. The damages are mainly due to abnormal machining force development during machining process [6]. The cutting force modeling techniques and state of the art are reported in Chapter 3.

Tool wear such as flank, crater, and notch wears, and built-up edge can have adverse effects on the surface finish of the machined components and can cause costly reworks. The surface and dimensional quality of parts are significantly influenced by the condition of cutting tools in the machining process. Tool fracture may lead to the scrapping of the machined parts and damages the machine tool components [39]. This can result in expensive equipment replacement, even bringing down the whole production line. To avoid cutting tools failures, they are often replaced before the end of their useful lifetime. It has been reported that only 50–80% of the expected tool life is typically used [40]. In the machining industry, 20% of downtime is attributed to tool failures [41]. The cutting tools wear is even more important when machining hard and brittle materials, that are in general characterized as *difficult to machine*. The processing of such materials can result in very high wear rates on both the flank and the face of the tool. In practice, the tooling cost, in the case of flexible manufacturing systems, represents approximately 25% of the total machining cost. In general tool life is characterized by randomness and its accurate prediction is quite difficult. The application of reliability techniques enable the calculation of tool life under process uncertainties [42]. There are several that have been investigated the reliability of cutting tools under different cutting conditions reported in chapter 5.

The present research studies applications of predictive analytics to the machining and cutting tools performance metrics such as cutting force, tool life and reliability and tool wear growth. Bayesian inference is used to develop probabilistic models to predict the performance metrics in turning and milling processes. In this context, the probabilistic models can predict the variables taking into account the inherent uncertainties of the process such as physical, model, measurement uncertainties.

Table 1.1: Comparison of the cutting force prediction with different modeling methods [43]

	Analytical	Numerical	Empirical	Probabilistic
Principle	Slip-line	Continuum mechanics using FEM, FDM, etc.	Curve fitting of experimental data	Probability theory
Capabilities	Predicts cutting forces, chip geometry, average stress, etc.	Predicts forces, chip geometry, stress, temperature, etc.	Applicable to most machining operations	Predicts cutting forces with the inherent uncertainty of the process
Limitations	Usually limited to 2D analysis	Material model, computational limitation: e.g., meshing	Valid only for the range of experimentation	Valid only for the range of experimentation
Advantages	Ability to develop fast, practical tools	Opportunity to connect to industrial relevant parameters	Practical, fast, and direct of industry-relevant parameters	Practical, requires few experimentations in the presence of informative prior knowledge, robust
Disadvantages	Unique to each machining problem	Long computational time	Extensive experimentation, time-consuming and costly	Difficulties with the prior selection, sometimes computationally expensive

1.2.1 Machining Process Modelling

Machining is the most general manufacturing operation regarding volume and expenditure. Machined components are found in almost every type of manufactured parts. It has been estimated that machining expenditure grants for approximately 5% of the GDP in developed countries. Over the last decades, challenges of the machining process have motivated researchers to predict the fundamental physical variables involved. However, the end goal of machining models is to predict industry-relevant outcomes and thus improve productivity. Table 1.1 summarizes modeling efforts for prediction of cutting forces in recent times [43].

1.2.1.1 Application of Soft Computing Modelling of Machining Process

Soft computing is an approach to computing resembling the considerable capability of the human mind to reason and learn in a domain of uncertainty and imprecision. In an attempt to discover reasonably useful solutions, soft computing-based techniques recognize the presence of imprecision and uncertainty in machining. Soft computing methods such as neural network (NN), fuzzy logic (FL), genetic algorithm (GA), simulated annealing (SA), ant colony optimization (ACO), and particle swarm optimization (PSO) have received remarkable attention due to their potentials to cope with nonlinear, multidimensional, and complex engineering problems [44].

Among the soft computing techniques, artificial neural networks (ANN) has been broadly used for modeling of the machining process. Neural networks are systems that can obtain, store, and utilize knowledge gained from experience. An ANN is an effective method of learning from an experimental dataset to describe the nonlinear and interaction effects between the variables, successfully. It is composed of an input layer used to incorporate data to the network, output layer to generate ANN's response and one or several hidden layers in between. The input and output layers interact with the environment, and hidden layers do not interact with the environment. NNs are identified by their topology, weight vectors, and activation function which are utilized in hidden and output layers of the networks. Neural networks are trained with some datasets and tested with different datasets to achieve optimal topology and weights. After the training procedure, the neural networks can be used for prediction [44].

Several researchers have proposed applications of NN to monitor and predict cutting force and tool life. Khanchustambham and *et al.* [45] used neural networks to predict cutting force and surface finish for machining of ceramic materials. Machining metrics such as feed, depth of cut, and spindle speed were used as input parameters of the network. The network was then trained using the cutting force signal and measured surface finish data for real-time monitoring of the turning process. Lee *et al.* [46] used fuzzy nonlinear programming combined with the neural network for prediction of cutting forces. The network is trained using cutting speed, depth of cut, and feed values as input parameters. The authors proved the effectiveness of the combined model, where the predicted results were in good agreement with experimental data. Dutta *et al.* [47] studied the application of neural network with different learning schemes using computers for faster data processing in online tool condition monitoring. The input variables of the neural networks were selected to be cutting speed, feed, depth of cut, and three forces components, and the output variable was flank wear. The authors reported that the modified back-propagation neural network (BPNN) has advantages over standard backpropagation algorithm regarding testing error level and rate of convergence. Quiza *et al.* [48] conducted an experimental investigation on the wear of ceramic cutting tools in turning of hardened cold rolled tool steel. They also predicted tool wear using neural networks and regression models, where the neural network model has advantages over the regression model.

1.2.1.2 Effect of Tool Geometry and Material on Cutting Tools Performance

The cutting edge geometry extremely influences cutting tools performance and life. A suitable shape of the cutting edge improves wear resistance, tool life, and process reliability. In this regard, the shape of the tools edge micro-geometries plays a vital role in reducing the cutting forces and increasing the process stability and tool life [49]. Endres et al. [50] investigated the effect of the tool edge and corner radius on orthogonal turning of AISI 1040. They investigated the influence of the tool geometry on the tool wear and cutting force, in which smaller edge and corner roundness led to less tool wear. Bassett et al. [51]

investigated the edge hone effect on tool life, thermomechanical stress, and cutting forces in the orthogonal cutting of the AISI 1045. Coated carbide inserts were selected, with a design micro-geometry having two different edge radii on the flank and rake faces. Denkena et al. [52] investigated the influence of cutting edge preparation on the performance of PVD coated tungsten carbide. They studied the effect of different geometries of the tool edge including triple edge chamfer, waterfall and trumpet geometries on the tool life and cutting forces during hard turning process. Wyen *et al.* [53] studied the effect of tool micro-geometry on cutting forces in machining of Titanium. The selected tool was tungsten carbide. They also reported the tool edge effect on the friction force, where the force was found to increase with an increase in the tool edge radius. Another comprehensive study of the tool geometry effect on the milling process performance was reported by Bouzakis *et al.* [54]. They investigated the edge preparation process for the coated tungsten carbide inserts and studied their influence on tool wear.

The researches mentioned above involve the investigation of the influence of tool micro-geometry on tool life. The present work aims to investigate the effect of cutting parameters and custom cutting tool geometries on the machining and cutting tool performance metrics using Bayesian-based predictive analytics, described in the next section.

1.2.2 Predictive Modelling for Design of Custom Cutting Tools

Figure 1.5 demonstrates the process of design and development for a custom cutting tool (with special geometry or material) in the industry. According to the figure, first, the prototype tool is designed and developed using the computer-aided engineering (CAE) software. Second, the tool is delivered to the testing department to conduct machining tests. Third, a test engineer performs machining experiment under defined cutting conditions to identify the performance of the custom tool. Forth, the collected information regarding the tool's performance such as tool life and wear is fed back to the tool designing department for further analysis.

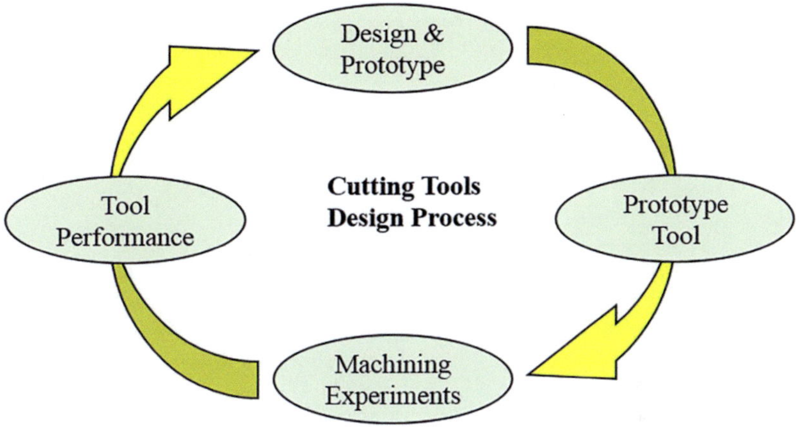

Figure 1.5: Information flow of cutting tool performance

Traditionally, cutting tools design and development, and the performance tests are conducted separately in industry. In this regard, the design phase is often implemented using FEA commercial software packages. Experiments then verify the results of the numerical analysis. The software packages are used to simulate stress distributing on the tool cutting edge, tool wear estimation, and prediction of the temperature of the chip-tool contact zone, among others. Notwithstanding their many benefits, there are several drawbacks to the programs, including their simplification of the machining and cutting tools parameters, as well as high computation cost and long run times. Figure 1.6 shows the functionality of the FEM software programs, input variables and predicting output variables.

The emergence of new sensors and monitoring equipment, and internet in modern machine tools ease the collection of the machining and cutting tools performance metrics. The data are recorded online or offline and fed back to the tool designer for modification of the design parameters. In this scenario, Bayesian inference can incorporate the tools performance data into the machining models in an efficient way. Moreover, the inherent uncertainties of the process reflected in the data can be quantified and minimized after each training steps of the model parameters.

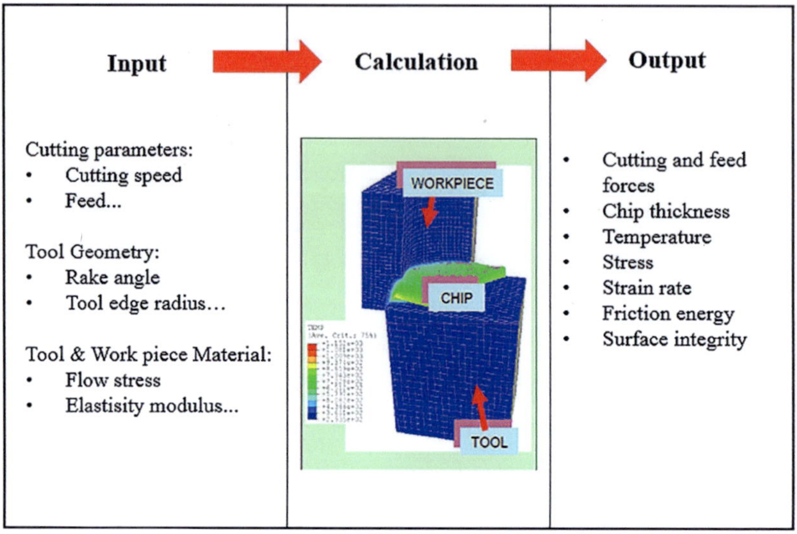

Figure 1.6: Functionality of the FEM software programs

Figure 1.7 demonstrate an architecture for Bayesian-based predictive analytics for the machining and cutting tools performance metric. As illustrated, Bayesian inference combines prior knowledge about a machining model parameter with the results of the experiments to predict the posterior knowledge of the model parameter. Using Bayesian inference, uncertainties of the prior knowledge can be minimized after training of the parameters with the machining experiments results (*e.g.*, cutting force and tool wear), so that the posterior probability is increased.

Bayesian inference can also be used as a sequential probabilistic approach for prediction of machining process variables. In this regard, the results of the posterior knowledge, *e.g.*, the first geometry can be integrated as a prior of the analysis for the second geometry. Bayesian inference is a knowledge-based approach which combines historical knowledge into the current and future analysis.

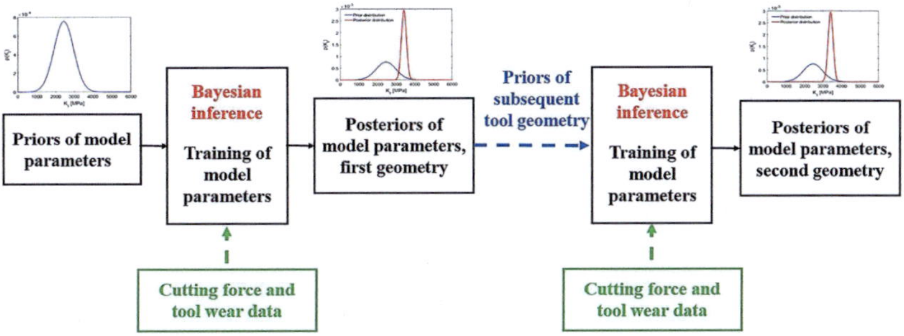

Figure 1.7: Bayesian-based predictive modeling for machining and cutting tools performance metrics

1.3 Research Objective

Advances in new sensors technologies and IoT enable physical devices to be connected and able to exchange data. Smart manufacturing systems such as smart machine tools can monitor the machining processes and detect the process anomaly before damages happen to the machine spindle or components. The rapid development of data analytics approaches has significantly improved the diagnostics and prognostics of the production systems. Data-driven analytics such as NN methods enable users to predict machining variables precisely. However, existing approaches and models applied to machining processes are often deterministic and therefore cannot model the inherent uncertainty of the machining process.

Moreover, they cannot integrate historical knowledge and expert opinion into current and future data analysis. The objective of this research is to investigate the applicability of Bayesian statistics for the design and manufacturing process as a predictive analytics approach. To demonstrate the advantage of using probabilistic method over the deterministic method, the Bayesian inference is applied to cutting forces, and cutting tools life, reliability, and degradation models. It is shown that Bayesian inference can be used for probabilistic modeling of the machining processes and design of the cutting tools under process uncertainties and using minimal experiments and inputs. The probabilistic modeling approach can predict the functionality, reliability, and probability of failure for cutting tools in turning and milling processes. This research also aims to demonstrate how the initial beliefs of the user or prior knowledge about a model parameter can be incorporated into the manufacturing data analysis and cutting tools design process. This avoids testing based on the null hypothesis and can minimize the number of often expensive and time-consuming experiments.

1.4 Overview of Later Chapters

This thesis is presented in three parts and seven chapters. The outline of the parts and the chapters are as follows. In part I, consisting of chapter 2, the fundamentals of Bayesian inference are presented, including prior's role in the Bayesian method, parameters estimation methods, MCMC method used to approximate posterior distributions, and application of MCMC to the bimodal distribution function, etc. In part II the application of Bayesian-based predictive analytics to the cutting force prediction is presented. In this part, Chapter 3 discusses applications of Bayesian inference to predict the cutting force using Merchant and Kienzle models, probabilistic cutting force prediction, and sequential modeling of the cutting forces for various cutting tools geometries. Chapter 4 discusses the development of extended Kienzle force model to isolate the ploughing force from the cutting force and probabilistic prediction of the forces. In part III applications of Bayesian-based predictive analytics to cutting tools life, reliability and tool wear growth analysis are reported. In this part, Chapter 5 discusses the tool life and cutting tools failure prediction, probabilistic sequential model development to predict the tool life, and the reliability analysis for different tool geometries. Chapter 6 covers the application of the Bayesian method of cutting tool degradation or tool wear growth prediction and sequential modeling approach to investigate the effect of tool geometry on tool wear growth. Finally, chapter 7 reports the conclusions and suggestions for future work.

2 Bayesian Inference

The probabilities are commonly used to express our observation and beliefs about unknown quantities informally. Nonetheless, the probabilities can be used to express information formally. In an accurate mathematical sense, it can be demonstrated that probabilities can numerically quantify a set of logical beliefs, that there is a relationship between probability and information. In this context, Bayes' rule provides a rational method for updating beliefs considering new information. The procedure of inductive learning using Bayes' rule is referred to as Bayesian inference [28].

Bayesian inference allows the prior or initial belief about a parameter to be updated by new observation(s). This approach offers a different view of hypothesis testing, compared to frequentist (classical) approaches. The Bayesian method enables users to integrate background knowledge into their analyses instead of testing the same null hypothesis repeatedly and ignoring the learnings from previous experiments. On the contrary, statistical methods based on the frequentist approach are often involved testing the null hypothesis. In other words, the null hypothesis implies that "no information is available." This hypothesis could be an inappropriate starting point for data analysis because it is almost always expected that "some information is available." Replication is an essential tool in design and manufacturing, and Bayesian methods fit within this framework because background knowledge can be incorporated into the statistical model. As a result, the possibility of previous research findings can be evaluated about new data, which makes the proposed approach an interesting tool for predictive analytics of the design and manufacturing applications [55].

In Bayesian inference, a probability is represented as a degree of belief. In this regard, uncertainty about a parameter can be quantified using a probability density function (PDF) and a cumulative distribution function (CDF). Bayesian inference is a powerful tool to quantify models, processes and measurement uncertainties, and to predict the parameter of interest with the probability of occurrence in various field of science, engineering, and economy, among others.

2.1 Bayes' Rule

Bayes' rule enables the prior, or initial belief about a parameter, to be updated by new experimental results, as shown in Eq. (2.1). According to the equation, the posterior probability, $p(\theta|y)$, is calculated by multiplying the prior probability, $p(\theta)$, by the likelihood function $p(y|\theta)$ and dividing by a normalizing function [28], as:

$$p(\theta|y) = \frac{p(y|\theta)\,p(\theta)}{\int p(y|\theta)\,p(\theta)\,d\theta} \tag{2.1}$$

The Bayesian approach to parameter estimation is to treat the parameter, θ, as a random variable. Considering Eq. (2.1), the prior distribution, $p(\theta)$, denotes how likely the parameter values are when no sampled data, y, is observed yet. In a Bayesian context, the aim is to estimate the posterior distribution, $p(\theta|y)$, over parameters given the sampled data. This is also known as posterior inference. Likelihood expresses the plausibility of a model parameter, θ, given specific observed data, y. Assuming independent and identically distributed (IID) observations, the likelihood function, $p(y|\theta)$, is written as the product of each observation probability shown in Eq. (2.2).

$$p(y|\theta) = p(y_1, y_2, y_3, \ldots, y_n|\theta) = \prod_{i=1}^{n} p(y_i|\theta) \tag{2.2}$$

The integral in Eq. (2.1) is the normalizing factor and is often referred to as the marginal probability. Generally, the integral does not have a closed-form solution, so it is convenient to simplify Eq. (2.1) to Eq. (2.3), [56], as:

$$p(\theta|y) = p(y|\theta)\,p(\theta) \tag{2.3}$$

Although integration of the posterior probability distribution, $p(\theta|y)$, is possible in some cases, in many cases, numerical approximations such as Markov Chain Monte Carlo (MCMC) methods is employed to generate samples from the posterior probability.

Figure 2.1 shows the process of updating the prior PDF by the likelihood function to achieve the posterior PDF.

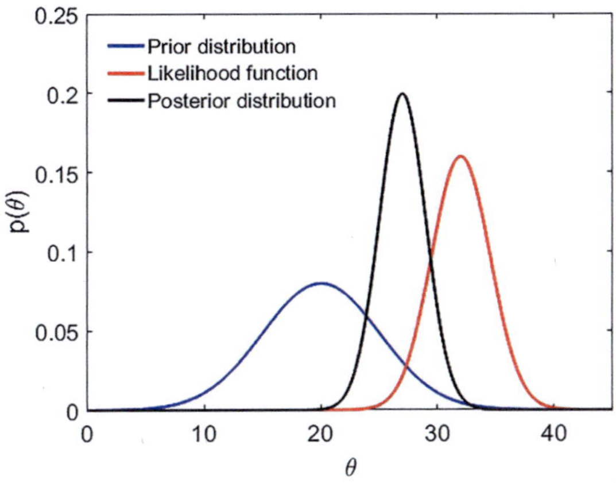

Figure 2.1: Schematic representation of the Bayes' rule

2.2 Role of the Prior in Parameter Inference

Prior distributions play an important role in parameter inference. They are essentially key parameters of Bayesian inference and represent the information about an uncertain parameter, θ. Different types of prior distribution exist, namely, informative and non-informative. Non-informative prior distribution has no population basis and plays a minimal role in the posterior distribution. The non-informative prior distribution is used when there is less degree of confidence about the parameter, θ, or large sample size is available. In this case, prior distribution of the parameter has minor effects on the posterior inference. The uniform distribution is frequently used as the non-informative prior distribution.

On the other hand, when the sample size is small, an informative prior has a stronger influence on the posterior distribution. An informative prior could come from operational or observational data, from previous experiments, or from engineering knowledge. In general, large sample sizes are required to modify strong priors, whereas weak priors are influenced by even relatively small sample sizes [57].

2.3 Comparison of Methods for Parameter Estimation

Every probability distribution has a set of parameters that need to be estimated. These parameters specify the constants are appearing in the model and provide a mechanism for efficient and accurate use of data [58,59]. There are some approaches to estimation of the parameters distributions, which are presented in the following sections [60].

2.3.1 Maximum Likelihood

Maximum Likelihood (ML) estimation method is a fundamental approach to estimate parameters. The objective of the method is to find the parameter estimates that maximize the likelihood of the observed data, y, given the model parameter, θ. This corresponds to:

$$\theta_{ML} = \arg max_{(\theta_i)}\ p(y|\theta_i) \tag{2.4}$$

where the θ_i vary across the range of allowable values. ML considers the parameter vector, θ, to be constant and provides maximum support for the evidence. Although this seems like a very straightforward means of estimating parameters in a probabilistic model, there are some serious drawbacks to this approach. For example, when we deal with small sample sizes, the ML estimate might not even be defined. It might also be nontrivial to find the maximum of the likelihood function when there are many parameters in the model [61].

2.3.2 Maximum a Posteriori

The posterior distribution is specified by a simple product of the likelihood and the prior. In Bayesian data analysis, one way to apply a model to data is to find the Maximum a Posteriori (MAP) parameter values. Using MAP, it is possible to find the parameter estimates that maximize the posterior probability of the parameter given the data. This corresponds to:

$$\theta_{MAP} = \arg max_{(\theta_i)}\ p(\theta_i|y) = \arg max_{\theta_i} p(y|\theta_i)\, p(\theta_i) \tag{2.5}$$

This looks similar to the ML estimation procedure. The difference is that the priors will influence the parameter estimation. Although it seems that the prior do not have important role in parameter estimation, there are several good reasons to prefer the MAP over ML estimation, especially when only a few data points are available [62].

2.3.3 Posterior Sampling: Markov Chain Monte Carlo (MCMC) Method

Although MAP method is useful for parameters estimates, particularly, when incorporating the prior belief to the estimation procedure, there are some drawbacks to the approach as follows:

1. MAP approach characterizes the posterior distribution with a single set of parameter values for a model. Nevertheless, a problem arises when there are multiple sets of parameter values that all have a very high posterior probability, or there are tradeoffs between parameter values. For example, suppose a model with two parameters A and B, where a high posterior probability is established either by setting A to a high value and B to a low value or the other way around. MAP estimate does not recognize such parameters correlations.
2. In a more comprehensive Bayesian method, the goal is to identify the full posterior distribution and not just to find the mode of the posterior distribution. In some cases, it could be possible to find an analytic expression for the posterior distribution. However, in many cases, it is required to resort to sampling techniques, such as MCMC, to get samples from the posterior distribution. These samples can be used to calculate some things, such as means, variances and other moments of the distribution. Using the sampling techniques, model parameters correlations could also be recognized.

MCMC technique is often applied to Bayesian inference and learning problems in large dimensional spaces. This technique plays a fundamental role in machine learning, physics, statistics, and econometrics, and decision analysis. A comprehensive literature review about MCMC approach has been reported by Andrieu *et al.* [63].

2.4 Bayesian Inference for Parameter Estimation

This section presents parameter estimation using probability distribution functions, via the normal and lognormal distributions. The functions are widely used for the representation of the cutting force and tool life data analysis and are adopted throughout this thesis. The Bayesian MCMC technique is applied to the problem of parameter estimation using a bimodal normal distribution.

2.4.1 Normal and Lognormal Distribution for Life Data Analysis

In probability theory, a normal (Gaussian or bell curve) distribution is the most common continuous probability distribution. It is particularly useful because of the central limit theorem. The central limit theorem establishes that, when independent random variables are added, their means (averages) tend toward a normal distribution even though the original variables themselves are not normally distributed [64]. There are two principal applications of the normal distribution to manufacturing engineering and reliability analysis investigated in this thesis. The first application deals with variability and uncertainty analysis of manufacturing process data. The second one concerns the lifecycle and failure analysis of the consumable items, such as cutting tools. The probability distribution functions of the normal lifetime distribution are PDF, CDF, reliability function, and hazard function. The normal lifetime PDF is expressed as follows,

$$f(t) = \frac{1}{\sigma\sqrt{2\pi}} e^{\frac{-(t-\mu)^2}{2\sigma^2}} \tag{2.6}$$

where μ is the mean, and σ is the corresponding standard deviation. Figure 2.2 displays the normal PDF lifetime distribution with mean value, μ, of 20 *min* and standard deviation, σ, of 3 *min*.

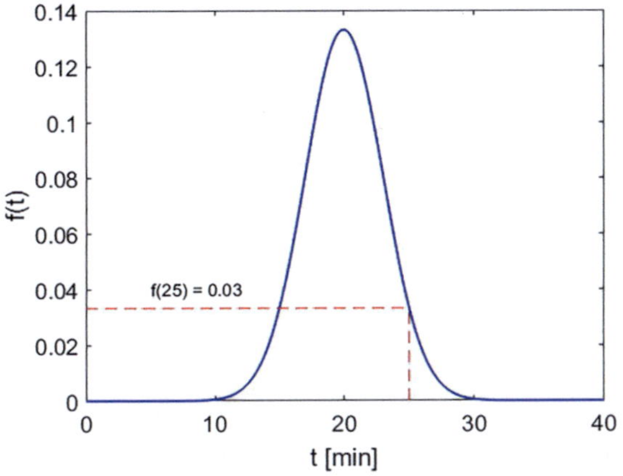

Figure 2.2: Normal PDF with a mean value of 20 *min* and a standard deviation of 3 *min*

In reliability study, lognormal distribution is commonly called lifetime distribution. The lognormal distribution is a continuous probability distribution of random variables whose logarithm is normally distributed. The lognormal PDF is extensively used to describe the distribution of positive random variables. A positive random variable, t, is lognormally distributed if the logarithm of t is normally distributed. This corresponds to:

$$\ln(t) \sim N(\mu, \sigma^2) \tag{2.7}$$

The PDF of the lognormal distribution is as follows:

$$f(t) = \frac{1}{t\sigma'\sqrt{2\pi}} e^{\frac{-(\ln t - \mu')^2}{2\sigma'^2}} ; t > 0 \tag{2.8}$$

where μ' and σ' are the mean and standard deviation of natural logarithms of the μ and σ. Figure 2.3 displays the lognormal lifetime PDF with a mean value of 20 *min* and a standard deviation of 3 *min*.

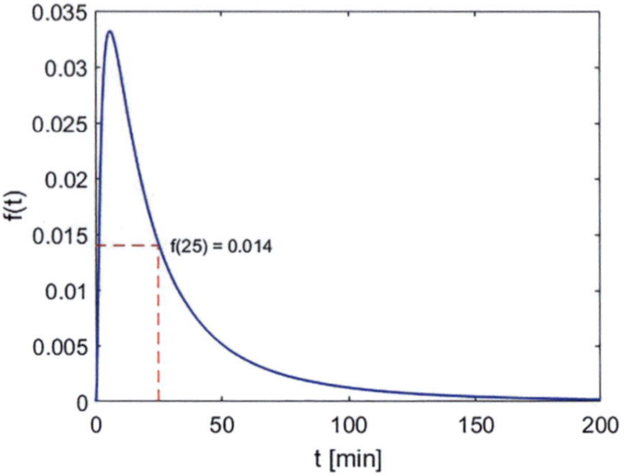

Figure 2.3: Lognormal PDF with a mean value of 20 *min* and a standard deviation of 3 *min*

2.4.2 Application of MCMC to Bimodal Normal Distribution

The MCMC method is a numerical approach to draw random samples, θ, from a distribution of interest. Metropolis-Hastings (MH) algorithm is the most widely used MCMC approach [63]. The MH technique can be used for drawing samples from symmetric and asymmetric proposal distributions. Metropolis algorithm is a special case of the MH algorithm, where the proposal function is symmetric. A normal distribution is often used as a symmetric proposal PDF denoted as $q(x)$. To illustrate the procedure, the Metropolis algorithm was used to approximate a known function (bimodal PDF) as the posterior target distribution; see Eq. (2.9).

$$p(\theta) \propto 0.4e^{(-0.2\theta^2)} + 0.6e^{(-0.2(\theta-10)^2)} \tag{2.9}$$

Algorithm 2.1: demonstrates the summary of the Metropolis algorithm for approximation of the posterior target distribution, $p(\theta)$. The algorithms allow to accept or reject the drawn samples from the proposal distribution, $q(\theta_{new}| \theta_i)$. In this regard, a new proposal, θ_{new}, is always accepted when it is more likely than the old state, θ_i. Therefore, the sampler moves towards the regions of the state space where the target function has a high density. In this example, the proposal distribution is symmetric normal distribution, such that $q(\theta_{new}| \theta_i) = q(\theta_i| \theta_{new})$. Therefore, the probability of proposing a new state given the old state is the same as proposing to go from the new state back to the old state. Cauchy, Student-t, and uniform distributions are other forms of the symmetric distributions, which can be used as proposal distributions of the Metropolis algorithm.

Algorithm 2.1: Metropolis algorithm for sampling of the posterior θ

1. Initialize a starting sample θ^0,
2. For i = 0 to i = N-1:
- Select a candidate θ^{new} from a proposal distribution, $q(\theta^{new}| \theta^i)$,
- Calculate the acceptance ratio,
  $r = \frac{p(\theta^{new})}{P(\theta^i)}$,
- Generate a random number,
  $u \sim uniform\ (0,1)$,
  If u ≤ r:
  Accept the proposal: $\theta^{i+1} = \theta^{new}$,
  Else:
  Reject the proposal: $\theta^i = \theta^{new}$,
  End If
3. End For

Figure 2.4 to Figure 2.9 display graphically the acceptance and rejecting procedure of the Metropolis sampling algorithms in six steps. In this context, Metropolis sampler is used to draw samples, θ_i, from the proposal distribution $q(\theta_{new}| \theta_i)$, to approximate the target function, $p(\theta)$ [60,63].

As illustrated in the Figure 2.4, θ_0 is selected as a starting point of the chain at the first step. At the second step, a candidate sample, θ_{new}, is drawn from the proposal distribution, $q(\theta_{new}| \theta_i)$; see Figure 2.5. In this case, the candidate sample is accepted, because the acceptance rate, r, is calculated (according to Algorithm 2.1) to be bigger than 1. Hence, the chain moves to θ_{new}.

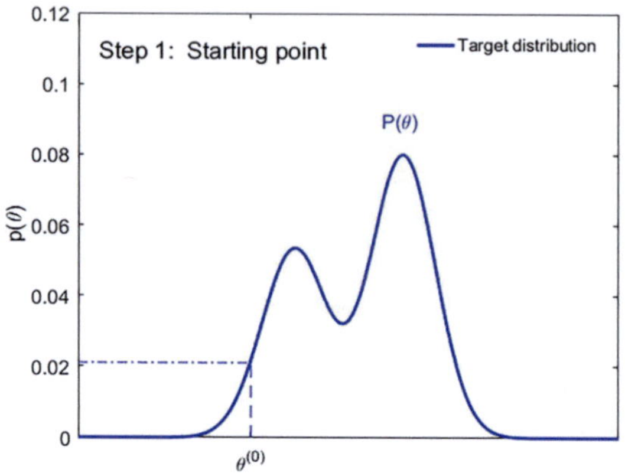

Figure 2.4: Sampling using the Metropolis algorithm. Step 1: starting point selection

At the third step, again a new candidate sample is selected from the proposal distribution, and then the posterior acceptance rate between the current and proposed sample is calculated; see Figure 2.6. In this case, the posterior rate, r, (ratio of the newly sampled parameter, θ_{new}, to the current sample, θ_1) is less than 1. The sample is not rejected but a new acceptance ratio, u, is generated randomly from a uniform distribution and compared with the posterior ratio, r. For this example, it is assumed that r is greater than u so that the sample is accepted, and the chain moves one step forward.

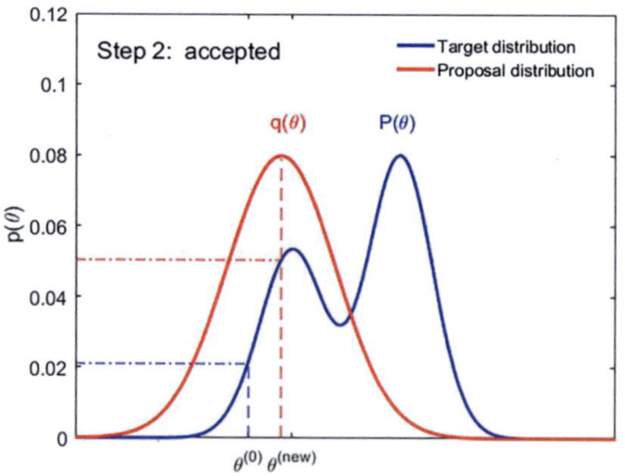

Figure 2.5: Sampling using the Metropolis algorithm. Step 2: accepted candidate sample

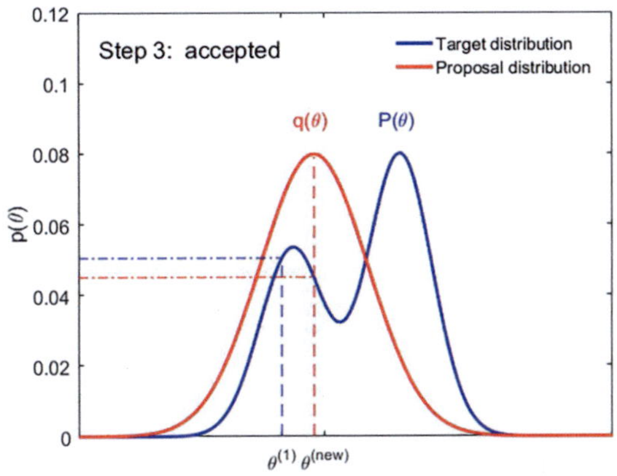

Figure 2.6: Sampling using the Metropolis algorithm. Step 3: accepted candidate sample

The algorithm is continued by proposing the fourth sample and calculating the acceptance ratios; see Figure 2.7. For this example, the sample is rejected, since both acceptance ratios are computed to be less and 1. Therefore, the chain remains at the current value of θ_2.

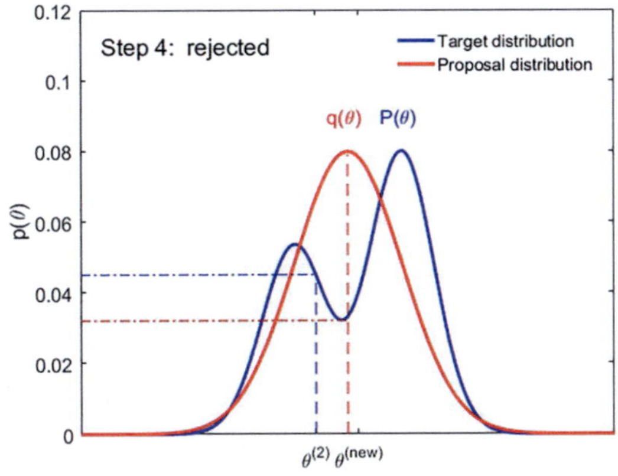

Figure 2.7: Sampling using the Metropolis algorithm. Step 4: rejected candidate sample

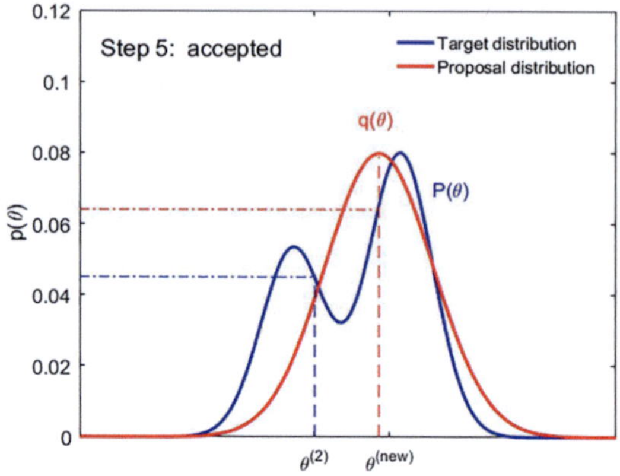

Figure 2.8: Sampling using the Metropolis algorithm. Step 5: accepted candidate sample

Figure 2.8 demonstrates the fifth step of the Metropolis algorithm iterations, where the newly drawn random sample from the proposal distribution, q, is accepted because the posterior ratio is greater than 1. As a result, the θ_3 value is assigned to the new sample θ_{new}. At the sixth step, the drawn proposed sample is rejected, because the posterior ratio is less than 1 and u; see Figure 2.9

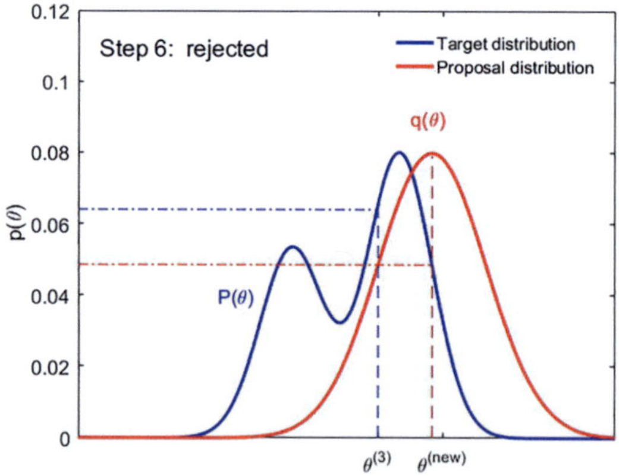

Figure 2.9: Sampling using the Metropolis algorithm. Step 5: rejected candidate sample

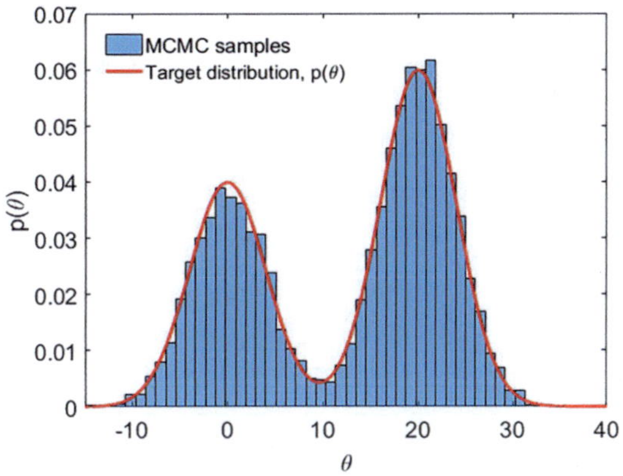

Figure 2.10: Approximation of the bimodal normal distribution using Metropolis algorithm

Figure 2.10 shows the histogram of the drawn samples and the normal bimodal function, $p(\theta)$. The Metropolis algorithm was practiced out for $N = 10{,}000$ iterations. According to the figure, the bimodal distribution in red color fits the histogram of the drawn samples shown in blue color. It is observed that the samples approximate the target PDF, $p(\theta)$, quite well. Note that the histogram and the target distribution were normalized to obtain a unit area under the curve.

2.4.3 Considerations Related to the MCMC Method

As demonstrated in the previous section, the key issue in the Metropolis algorithm is the acceptance or rejection of the current state in the Markov chain. In order for the chain to explore the solution space more efficiently, it is essential for samples to be accepted in each stage. Poor algorithms reject samples often and force very long runs to achieve steady state condition and an appropriate sample of the possible states. The second issue with the MCMC method is the convergence of the Markov chain. In order to ensure convergence, the burn-in technique is used, which is meant to give the Markov Chain time to reach its stationary distribution. The technique is referred to as discarding the first n samples of the iterations. The idea is that a "bad" starting point may oversample regions that are very low probability under the equilibrium distribution before it settles into the stationary distribution. A practical way to evaluate the convergence of the chain's stationary distribution (appropriate length of the burn-in period) is by observing the trace plot and histogram of the parameters [65]. The third issue with the Metropolis simulation is that the values of the sampled parameters are correlated since they are generated by the Markov process. Excessive autocorrelation may cause a problem with the model specification and should be investigated further. To reduce the autocorrelation, there are two techniques, which are explained as follows,

1. **Thinning**: This refers to drawing samples in a regular interval. For instance, in the previous section, $N = 10{,}000$ iterations were exercised, where each iteration was repeated for 10 times and every 10th value was kept.
2. **Tuning the proposal distribution:** A key factor in achieving high efficiency in Markov chain is finding a suitable proposal distribution for each parameter. The tuning can be examined by acceptance probability of a sampling process. This refers to the percentage of the proposals that have been accepted. A high acceptance rate means that most new samples occur right around the current data point. This indicates that the Markov chain is not fully exploring the parameter space. On the other hand, a low acceptance rate means that the proposed samples are often rejected; hence the chain does not move much forward. Roberts and Rosenthal [66] demonstrated empirically that an acceptance rate between 0.15 and 0.50 is at least 80% efficient. Moreover, Hoff [67] stated that to improve the performance of the Markov chain, the posterior variance of samples can be an efficient choice of the proposal variance.

To illustrate the evaluation of the chain's convergence and reduction of the autocorrelation, two case studies are presented. In the first case study, the proposal variances of the parameter, θ, is selected to be 0.3; while in the second scenario, the variance is 13. In both cases, the number of the iterations and burn-in samples are selected to be 1×10^4, and 1000, respectively. Using the variance of the first case study, the trace plot of the burn-in period shows that the chain cannot converge to a stationary condition, as can be seen in Figure 2.11. In the example discussed above, the histogram of the drawn sample cannot approximate the known distribution function (bimodal normal distribution), even though a very high acceptance ratio of 0.977 is achieved; see Figure 2.12.

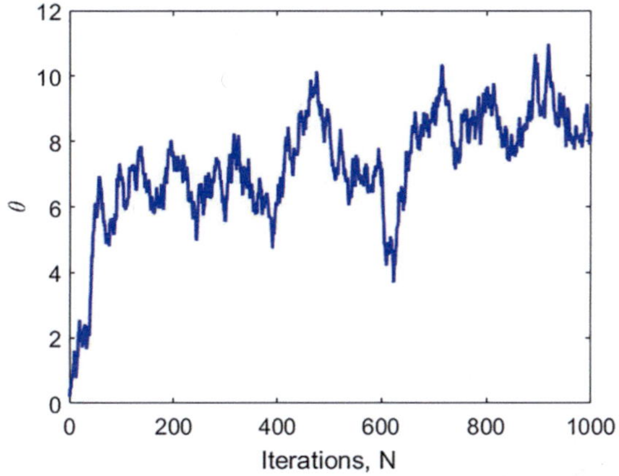

Figure 2.11: Trace plot of the 1000 initial samples drawn from the θ target distribution, un-stationary chain

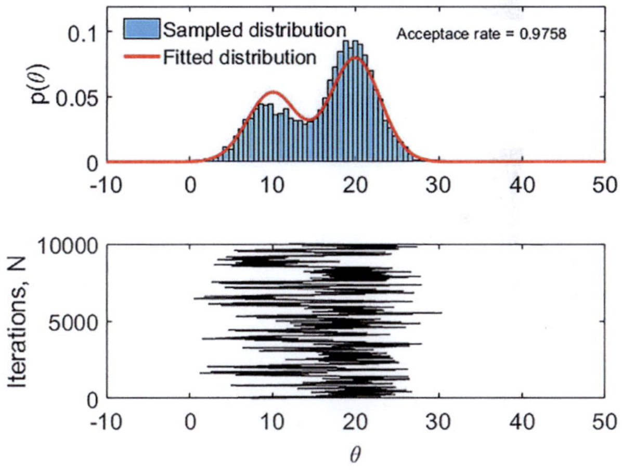

Figure 2.12: High acceptance ratio as a result of the proposal variance 0.3

Figure 2.13 shows the autocorrelation of the first and last 100 samples of the parameter, θ, after discarding the burn-in period. According to the figure, the high autocorrelation of 0.4 is seen at the lag of 50 for both the first and last 100 samples.

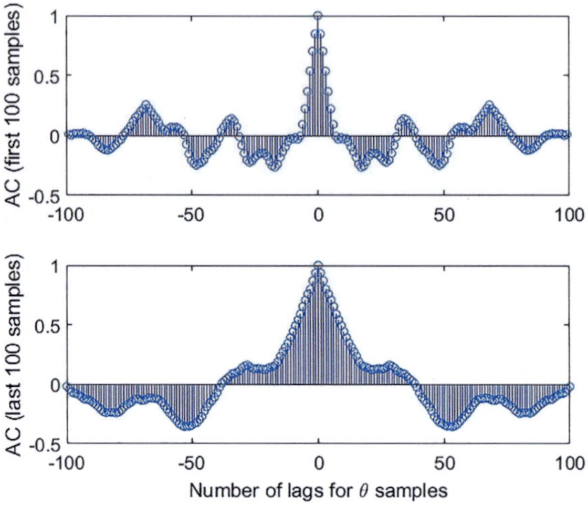

Figure 2.13: Autocorrelation of first and last 100 samples, choosing proposal variance 0.3

Nevertheless, using the variance of the second case study, the trace plot of the burn-in period shows that the chain converges to a stationary condition see Figure 2.14. Furthermore, the histogram of the drawn sample approximates the known distribution function (bimodal normal distribution) precisely; see Figure 2.15. As can be seen, the obtained acceptance ratio is 0.445.

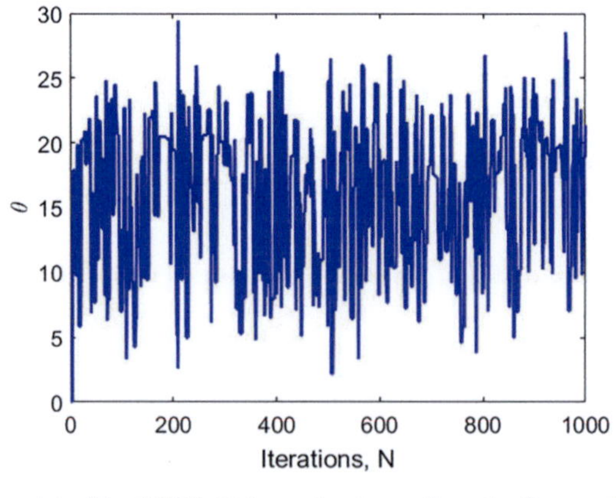

Figure 2.14: Trace plot of the 1000 initial samples drawn from the θ target distribution, stationary chain

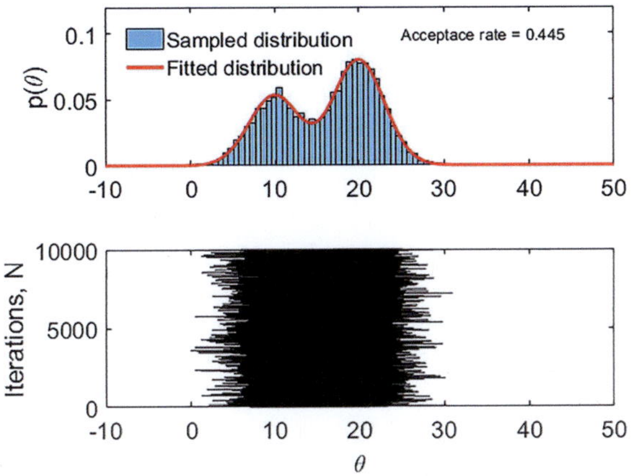

Figure 2.15: Suitable acceptance ratio of 0.445 because of the proposal variance 13

Figure 2.16 displays the autocorrelation of first and last 100 samples of the parameter θ after discarding burn-in period. According to the figure, the autocorrelations were minimized to less than of 0.2 for the entire range of the samples. Comparing two case studies, it was shown that the second proposal variance leads to better convergence of the trace plot as well as less correlated samples in the chain.

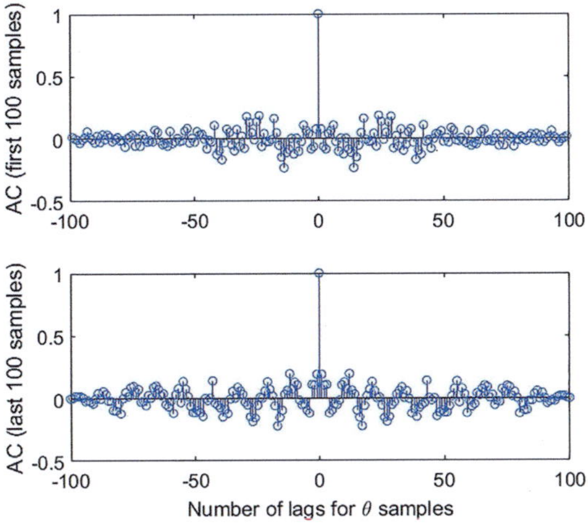

Figure 2.16: Autocorrelation of first and last 100 samples, choosing proposal variance 13

2.4.3.1 Convergence Diagnostic for Markov Chains

Geweke [68] proposed a convergence diagnostic for Markov chains based on a comparison of the last part of the chain against some smaller interval at the beginning of the chain (*e.g.*, the first 10% and last 50% after removing the burn-in period). If the chain is at the stationary condition, the means of the samples are almost equal. The Geweke test is applied to ensure the convergence of the Markov chain for all the simulations throughout this thesis.

2.4.3.2 Sampling Multivariate Distributions

In the previous example on Metropolis MCMC method, the focus was on how to sample from the univariate target distribution (bimodal PDF). This was performed to give readers some intuition for implementation of MCMC with an example that can be visualized. Nevertheless, the Metropolis MCMC can be used to sample multivariate distributions. There are two procedures for sampling distributions in multiple dimensions, which are described as following [60,69]:

1. **Blockwise method:** In this approach, proposal distribution, $q(\Theta)$ is selected to have the same dimensionality as the target distribution, $p(\Theta)$. To demonstrate, consider $p(\Theta)$ as a target PDF of n variables (*i.e.*, $\Theta = (\theta_1, \theta_2,\ldots, \theta_n)$) and $q(\Theta)$ as the proposal distribution having identical dimensionality with the $p(\Theta)$. The proposed state, Θ_{new}, which is sampled from $q(\Theta)$, is either accepted or rejected in the same way as for the univariate Metropolis algorithm.
2. **Componentwise random walk method:** In this method, the algorithm samples one variable at a time and then proceeds sequentially to sample the remaining variables of a multivariate PDF in the order of $\theta_1 \rightarrow \theta_2 \rightarrow \theta_3 \rightarrow \ldots \theta_n$. The sampling for each variable is conducted using a univariate proposal distribution, $q(\theta_i)$, for that variable, θ_i. This method is particularly useful when the number of variables, n, becomes large.

In this research, blockwise updating technique of the Metropolis MCMC is used to sample from multivariate probability PDFs. Additionally, the proposal distributions are taken to be normal distribution in all instances.

Part II: Bayesian-Based Analytics for Cutting Force Prediction

3 Bayesian Updating for Sequential Cutting Force Prediction in Turning Process[1]

3.1 Introduction

Machining models provide relationships between user-selected inputs (feed, cutting speed, tool geometry) and process outputs, such as cutting forces. The models may be numerical or analytical in format. In this context, several models have been proposed [70–73] to predict cutting forces in milling and turning operations. In general, machining models are deterministic. In other words, given a set of inputs, a unique set of outputs is obtained. However, to establish a predictive model, the mean and distribution in the outputs must be related to the input means and distributions. This probabilistic approach incorporates the inherent uncertainties. These uncertainties are due to the machine and machining process, workpiece material, measurement process, tool material, and tool geometry, among others. The uncertainty evaluation and probabilistic prediction of cutting force can be performed by Bayesian inference [65]. Schmitz *et al.* [74] investigated cutting force prediction under uncertainty using Bayesian inference for the Merchant model, where the cutting force is linearly proportional to the feed-dependent uncut chip thickness. Discrete grid method was used to update the force model parameters. Mehta *et al.* [75] developed a mechanistic force model for cutting force prediction using Markov Chain Monte Carlo approach again applied to the Merchant model. Gözü and Karpat [76] studied the application of Bayesian inference to predict cutting forces in micromilling of Titanium alloy TiAl4V. The Metropolis-Hasting algorithm of MCMC was used to identify probability distributions of the cutting and

[1] This chapter is extended from the following publication: M. Salehi, T.L. Schmitz, R. Copenhaver, R. Haas, J. Ovtcharova, Probabilistic Sequential Prediction of Cutting Force Using Kienzle Model in Orthogonal Turning Process, J. Manuf. Sci. Eng. 141 (2018) 11009. doi:10.1115/1.4041710.,

ploughing forces coefficients based on experimental measurements and mechanistic models of micromilling. The mechanistic model can predict the cutting and ploughing forces in radial and tangential directions, where the cutting forces are linearly proportional to the uncut chip thickness.

The force models, which have been used in the papers mentioned above describe linear relationships between the feed values and the cutting and ploughing forces. Nevertheless, the probabilistic cutting force prediction considering the *size effect* phenomenon is yet to be investigated using nonlinear models such as Kienzle force model [77]. The term size effect refers to as the nonlinear increase of the specific cutting energy with decreasing the undeform chip thickness. A good summary of the phenomenon and the modeling techniques are given by Vollersten *et al.* [78]. In this chapter, the Bayesian inference is applied to the Merchant and Kienzle force models to predict the cutting forces at very low feed values in turning. Metropolis algorithm of the Markov Chain Monte Carlo (MCMC) method is used to estimate the force models' parameters. In order to investigate the effect of cutting tool geometry on cutting forces, two cutting tool chamfer (rake) angles, 0 and -10 *deg*, are tested under different cutting conditions in an orthogonal turning process. First, the probabilistic prediction of tangential force using the Merchant model for the 0 *deg* rake angle tool is presented. Next, the Bayesian inference is applied to the Kienzle model to predict tangential and feed forces for the 0 and -10 *deg* rake angle tools, sequentially.

The chapter is organized as follows. In section 3.2, deterministic models of Merchant and Kienzle are presented. In section 3.3 the experimental setup and the force measurement results are presented. In section 3.4, the Bayesian inference scheme and MCMC method are presented. Section 3.5 describes the application of MCMC to the Merchant using the 0 *deg* tool rake angle. Section 3.6 presents the application of MCMC to the Kienzle force models using the 0 *deg* tool rake angle. The results of the posterior forces and parameters of the Kienzle model for the 0 *deg* tool rake angle are used to predict forces for the -10 *deg* rake angle tool in section 3.7. Conclusions are provided in section 3.8.

3.2 Deterministic Cutting Force Models

Deterministic models of the Merchant and Kienzle force models are presented in this section. The Mechanistic Merchant model is based on the assumption that the tool edge radius is zero. However, Kienzle force model takes into account the effect of cutting edge radius on specific cutting force coefficient. Weber *et al.* [77] investigated the increase of the force coefficients comparing two edge radii of 5 and 50 μm, where the larger cutting edge radius resulted in a higher specific cutting force. They also reported that the edge radius of the tool used in the investigations to derive the Kienzle equation was probably in the range of 10–20 μm.

3.2.1 Merchant Force Model

Merchant force model describes linear relationships between cutting force and the uncut chip thickness, h [38]. Figure 3.1 displays a schematic orthogonal cutting model for the cutting forces calculation.

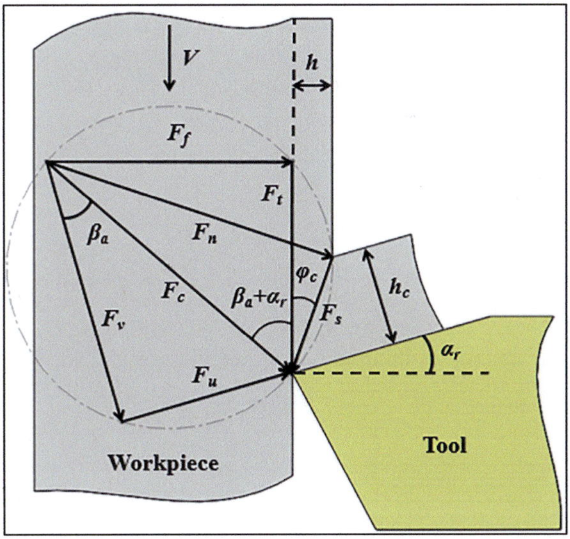

Figure 3.1: Merchant cutting force diagram for a negative rake angle tool

According to the figure, the tangential force F_t and the feed force F_f are achieved as below,

$$F_t = K_t bh \tag{3.1}$$

$$F_f = K_f bh \tag{3.2}$$

where K_t and K_f are cutting force coefficients, b is the width of cut, and h is the uncut chip thickness. The width of cut and the feed value are decided by the machinist, while the cutting force coefficients must be calculated. In order to calculate F_t and F_f, one needs to find K_t and K_f as follows,

$$K_t = \tau_s \frac{cos(\beta_a - \alpha_r)}{sin(\emptyset_c)\, cos(\emptyset_c + \beta_a - \alpha_r)} \tag{3.3}$$

$$K_f = \tau_s \frac{sin(\beta_a - \alpha_r)}{sin(\emptyset_c)\, cos(\emptyset_c + \beta_a - \alpha_r)} \tag{3.4}$$

where τ_s is the shear stress along the shear plane, ϕ_c is the shear plane angle, β_a is the average friction angle, and α_r is the tool rake angle. The τ_s is determined through the equations (3.5) to (3.7) as follows,

$$\tau_s = \frac{F_s}{A_s} \tag{3.5}$$

$$F_s = F_t\, cos(\emptyset_c) - F_f\, sin(\emptyset_c) \tag{3.6}$$

$$A_s = \frac{bh}{sin(\emptyset_c)} \tag{3.7}$$

where the F_s is the shear force and A_s is the shear area. The β_a is achieved as follows,

$$\beta_a = \alpha_r + tan^{-1}(\frac{F_f}{F_t}) \tag{3.8}$$

Shear plane angle, ϕ_c is obtained as follows,

$$\emptyset_c = tan^{-1}(\frac{r_c\, cos(\alpha_r)}{1 - r_c\; sin(\alpha_r)}) \tag{3.9}$$

$$r_c = \frac{h}{h_c} \tag{3.10}$$

where h_c is the cut chip thickness, and r_c is the chip thickness ratio.

3.2.2 Kienzle Force Model

The Kienzle force models, Eq. (3.11) and (3.12) describe nonlinear relationships between the uncut chip thickness, h, and the cutting force components in tangential and feed directions, F_t and F_f, respectively [79]:

$$F_t = K_{tt}.b.h^{1-c_t} \tag{3.11}$$

$$F_f = K_{ff}.b.h^{1-c_f} \tag{3.12}$$

where 1-c_t and 1-c_f exponents are positive constants less than one; while, K_{tt} and K_{ff} are tangential and feed specific cutting force coefficients. The force coefficients depend on the workpiece material, and the exponents depend on the geometrical cutting parameters, cutting speed and the tool-workpiece combination.

3.3 Experimental Setup, Results, and Discussion

Tube turning experiments were performed on a Haas TL-1 CNC lathe; see Figure 3.2. The dry machining tests were completed using a Kennametal turning insert, CCMW3252, with 0 *deg* rake angle as well as uncoated inserts SPGW09T308 with the rake angle -10 *deg*. The latter inserts were designed and produced with the special edge geometry by Zermet Zerspanung GmbH, with the ISO grade of P25. The tubular workpiece material was 1020 steel with an outer diameter of 25.4 mm and wall thickness of 2.1 mm.

Figure 3.2: Machining experiments setup

The corresponding chip width was 2.1 mm. Feed values of f = {0.051, 0,076, and 0,102} *mm/rev*, as well as three cutting speeds of V_c = {60, 80 and 100} *m/min*, were selected. The experiments were repeated three times for each cutting speed-feed combination. Therefore, the total number of experiments was 54.

Before starting the machining tests, micro geometry of the special tools was inspected using a laser scanning microscope, Keyence VK8710; see Figure 3.3. Figure 3.4 shows the result of the tool chamfer (rake) angle measurement, where the exact rake angle was measured to be -10.62 *deg*. The measurement process was repeated for three samples to ensure the precision and the repeatability of the edge preparation process. According to the results, the mean and standard deviation of the rake angle were -10.10 and 0.56 *deg*, respectively.

Surface roughness was also measured along the cutting edge for three sample tools; see Figure 3.5. According to the results, mean and standard deviation of the edges roughness based on ten-point mean roughness (R_z) were achieved to be 7.5 and 0.36 μm, respectively. Additionally, the mean and standard deviation of cutting edges based on arithmetical mean roughness (R_a) were obtained to be 1 and 0.13 μm, respectively.

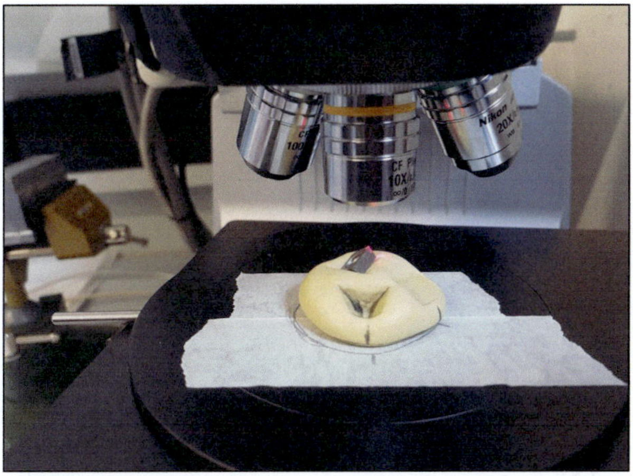

Figure 3.3: Tool geometry inspection with Keyence laser scanning microscope

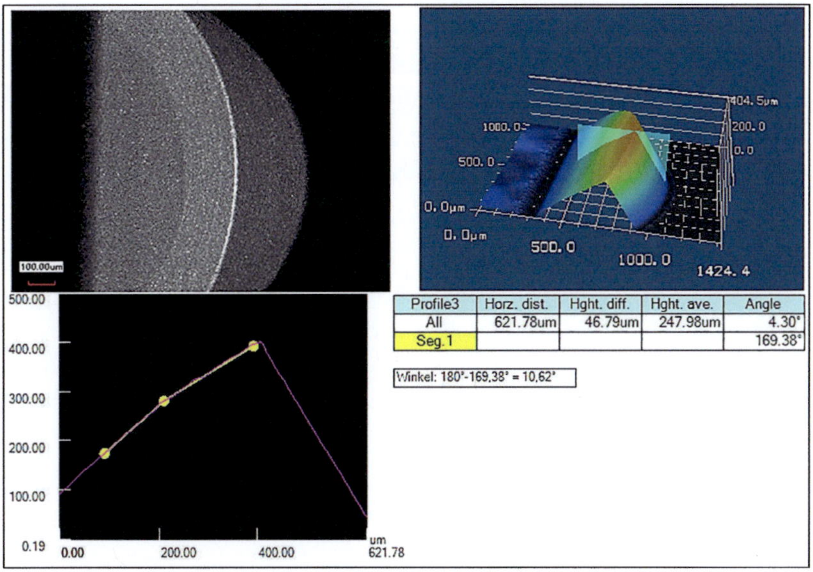

Figure 3.4: Measurement of the tool rake angle with Keyence laser scanning microscope

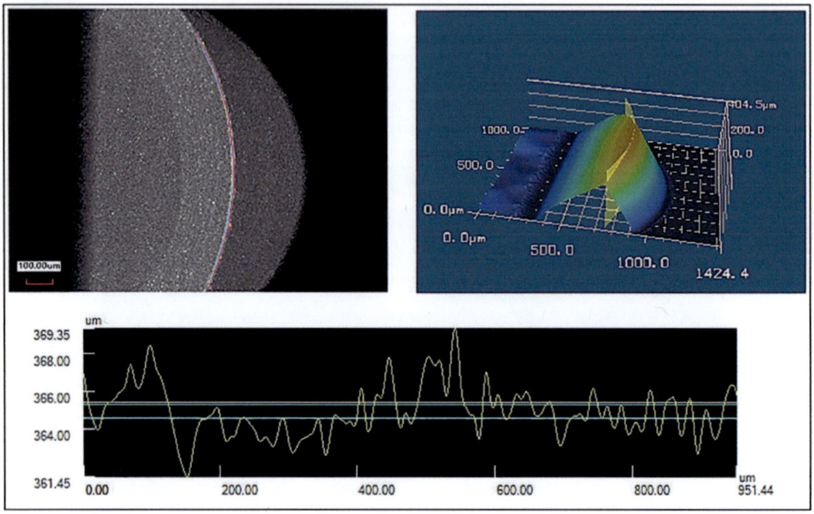

Figure 3.5: Measurement of the tool edge roughness with Keyence laser scanning microscope

A three-axis force dynamometer (Kistler 9257B) was used to measure the cutting force in tangential and feed directions. Machining chips were collected to measure the thickness by dial caliper. For each geometry, three data sets were selected to update the prior of the probabilistic models, while the others were used for model verification.

Digital high-speed camera (Fastec IL-3) with a maximum frame rate of 1250 *frames/sec*, was used to record videos from the cutting zone. Matlab image processing toolbox was utilized to obtain frame by frame images from the recorded videos, and determination of shear plane angle with the corresponding cut chip thickness. Since the obtained images correspond to the operations under various cutting speeds and feed values at different frames, various shear plane angles were observed. Figure 3.6 shows the shear plane angles quantities using tool rake angle 0 *deg*, under different cutting data.

The results of the observation were tabulated in Table 3.1. The mean value and standard deviation of the shear plane angles were computed, 17.7, and 4 *deg*, respectively. The values are used as an initial belief of the shear plane angle for Bayesian modeling.

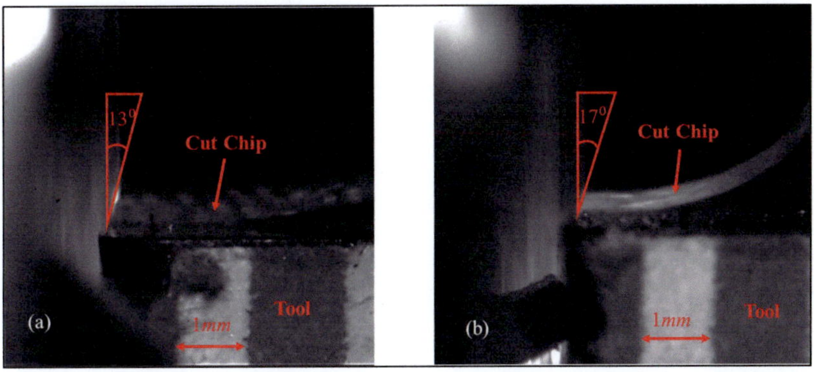

Figure 3.6: Cutting zone images to determine the shear plane angle using tool rake angle 0 *deg*, cutting feed 0.102 *mm/rev*, and the cutting speeds, a) 60 *m/min*, b) 80 *m/min*, and

Table 3.1: Measured shear plane angles using high-speed camera images for the tool rake angle 0 *deg*

No.	V_c *(m/min)*	*f (mm/rev)*	h_c *(mm)*	ϕ_c *(deg)*
1	60	0.051	0.24	15
2	60	0.076	0.25	16.5
3	60	0.102	0.33	13
4	80	0.102	0.3	17
5	80	0.102	0.28	19
6	100	0.102	0.22	22
7	100	0.102	0.2	25

From the maximum shear stress principle in addition to Lee and Shaffer's slip line model [38], average friction angle was achieved, 27.3 *deg*, (see Eq. (3.13)). The value can be used as an initial belief of the average friction angle.

$$\beta_a = \frac{\pi}{4} - (\phi_c - \alpha_r) \tag{3.13}$$

3.3.1 Cutting Force and Chip Thickness Measurement

Twelve tangential and feed forces were selected to be used for training of the prior model parameters. In addition, chip thickness of the corresponding experiments was measured by dial caliper in various points along the chip length, so that the variations in the thickness were obtained and reported as standard deviation values.

3.3.1.1 Cutting Forces for Training of the Model Parameters

Figure 3.7 and Figure 3.8 display the tangential and feed force component data under different cutting conditions, using the tool rake angle 0, and -10, *deg*, respectively.

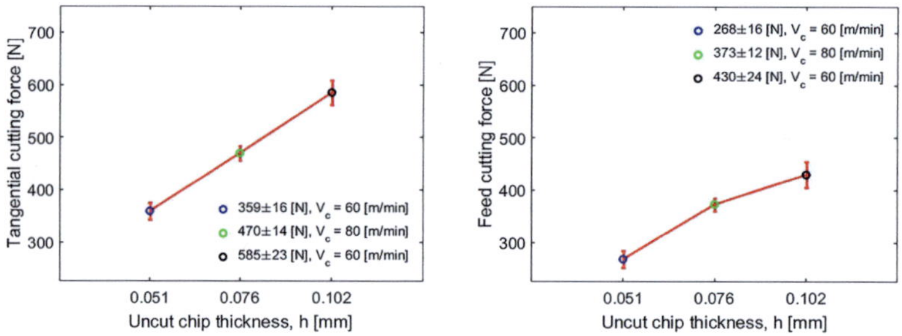

Figure 3.7: Tangential and feed force components for training of prior using tool rake angle 0 *deg*

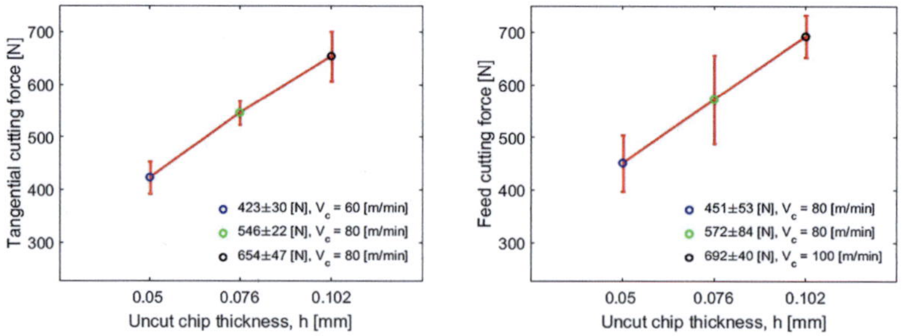

Figure 3.8: Tangential and feed forces for training of prior using tool rake angle -10 *deg*

The mean is provided together with one standard deviation error bars. As can be seen in the figures, the forces increase with an increase in feed for the all geometries; while the cutting speed has a slight influence on the cutting forces. Using tool rake angle 0 *deg*, feed forces values are seen less than tangential forces; nevertheless, the feed force component goes beyond the tangential component for the bigger negative rake angles -10 *deg*.

3.3.1.2 Chip Thickness for Training of the Merchant Model Prior

Figure 3.9 shows the mean and one standard deviation of cut chip thicknesses. The values are used to train the priors of the merchant model parameters for the 0 *deg* rake angle tool. According to the figure, the cut chip thickness increases with an increase in feed values.

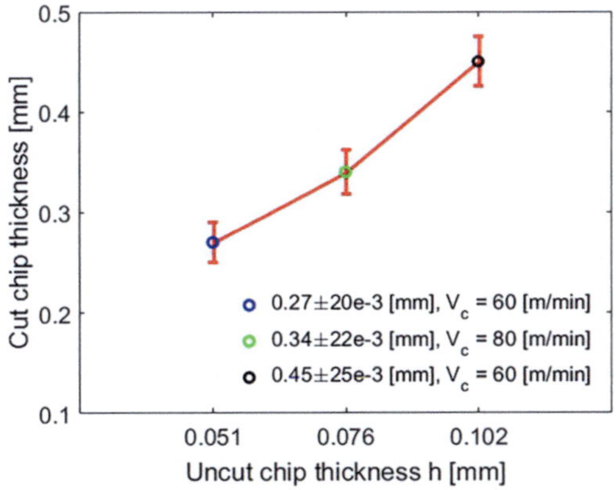

Figure 3.9: Mean and standard deviation values of chip thickness using tool rake angle 0 *deg*

3.4 Metropolis Algorithm for Merchant and Kienzle Models

Bayesian MCMC method is used to estimate Merchant and Kienzle model parameters and quantify the corresponding uncertainties. To demonstrate, Metropolis algorithm of MCMC is used to update Merchant and Kienzle force model parameters; see Algorithm 3.1 and Algorithm 3.2.

Algorithm 3.1: Metropolis algorithm for updating the Merchant force model parameters

1. Establish a normal prior distribution, $p(\phi_c)$,
2. Establish a proposal density function for ϕ_c,
3. Initialize a starting sample ϕ_c^0,
4. For i = 0 to i = N-1:
- Select a candidate ϕ_c new from a proposal distribution, $q(\phi_c^{new}| \phi_c^i)$,
- Compute the posterior distribution,
  $p(\phi_c{}^i| \phi_c{}^m) = p(\phi_c{}^i)\, p(\phi_c{}^m| \phi_c{}^i)$,
- Calculate the acceptance ratio,
  $r = \frac{p(\phi_c{}^{new})}{P(\phi_c{}^i)}$,
- Generate a random number,
  $u \sim uniform\ (0,1)$,
  If $u \leq r$:
  Accept the proposal: $\phi_c{}^{i+1} = \phi_c{}^{new}$,
  Else:
  Reject the proposal: $\phi_c{}^i = \phi_c{}^{new}$,
  End If
5. End For
6. Establish a normal prior distribution, $p(\beta_a,\tau_s)$,
7. Establish a proposal density function for (β_a,τ_s),
8. Initialize a starting sample $(\beta_a,\tau_s)^0$,
9. For i = 0 to i = N-1:
- Select a candidate $(\beta_a,\ \tau_s)^{new}$ from a proposal distribution,
  $q((\beta_a,\tau_s)^{new}| (\beta_a,\tau_s)^i)$,
- Compute the posterior distribution,
  $p((\beta_a,\tau_s)^i| F_c{}^m,\phi_c{}^i) = p((\beta_a,\tau_s)^i)\, p(F_c{}^m| \phi_c{}^i,(\beta_a,\tau_s)^i)$,
- Calculate the acceptance ratio,
  $r = \frac{p((\beta_a,\tau_s)^{new})}{P((\beta_a,\tau_s)^i)}$,
- Generate a random number,
  $u \sim uniform\ (0,1)$,
  If $u \leq r$:
  Accept the proposal: $(\beta_a,\tau_s)^{i+1} = (\beta_a,\tau_s)^{new}$,
  Else:
  Reject the proposal: $(\beta_a,\tau_s)^i = (\beta_a,\tau_s)^{new}$,
  End If

10. End For

Algorithm 3.2: Metropolis algorithm for updating of the Kienzle force model parameters

1. Establish a normal prior distribution, $p(K_{tt}, c_t)$,
2. Establish a proposal density function for (K_{tt}, c_t),
3. Initialize a starting sample $(K_{tt}, c_t)^0$,
4. For i = 0 to i = N-1:
- Select a candidate $(K_{tt}, c_t)^{new}$ from a proposal distribution, $q((K_{tt}, c_t)^{new} \mid (K_{tt}, c_t)^i)$,
- Compute the posterior distribution, $p((K_{tt}, c_t)^i \mid F_t) = p((K_{tt}, c_t)^i)\, p(F_t \mid (K_{tt}, c_t)^i)$,
- Calculate the acceptance ratio, $r = \frac{p((K_{tt}, c_t)^{new})}{P((K_{tt}, c_t)^i)}$,
- Generate a random number, $u \sim uniform\ (0,1)$,

  If $u \leq r$:

    Accept the proposal: $(K_{tt}, c_t)^{i+1} = (K_{tt}, c_t)^{new}$,

  Else:

    Reject the proposal: $(K_{tt}, c_t)^i = (K_{tt}, c_t)^{new}$,

  End If

5. End For

According to the Algorithm 3.1, Metropolis algorithm is applied to approximate the posterior target distribution of the parameter ϕ_c, first. Second, the algorithm is used to sample from the joint posterior distribution of the parameters (β_a, τ_s). The likelihood function in the first algorithm calculates the probability of the measured shear plane angle, $\phi_c{}^m$, given the shear plane angle, ϕ_c. The likelihood function in the second algorithm calculates the probability of the measured force, F_c, given the model parameters, ϕ_c, and (β_a, τ_s). It is important to note that, the Metropolis algorithm is able to accept or reject the candidate samples of the model parameters depending on the acceptance ratio of the posterior probabilities.

Algorithm 3.2 is applied to draw samples from the joint posterior target distribution, $p(K_{tt}, c_t)$. According to the algorithm, the likelihood function calculates the probability of the measured tangential force, F_t, given the model parameters (K_{tt}, c_t). The posterior distribution is calculated by multiplying the prior joint distribution into the likelihood function.

To reduce the excessive autocorrelation of the drawn samples using the Metropolis algorithm, the thinning technique is performed. Additionally, the proposal distribution of the samples is tuned by selecting the sample acceptance ratio roughly between 15-50 % [56,66]. Geweck's method [68] is used for the convergent diagnostics of the drawn sample of the posterior target function.

3.5 Application of MCMC to Merchant Force Model using Tool Rake Angle 0 *deg*

In the Merchant force model, there is uncertainty in the force coefficient, K_t, due to the uncertainty in the model parameters, ϕ_c, β_a, and τ_s. The uncertainty evaluation and minimization using Bayesian MCMC are explained in this section. Summary of the steps of the MCMC application to the force models is described as follows,

1. Establishing the priors of the force models' parameters.
2. Parameters updating using the likelihood function of the measured forces.
3. Computing of the posterior distribution of the force models' parameters and cutting forces using MCMC Metropolis algorithm.

3.5.1 Establishing the Prior Distributions

Priors of the model parameters were obtained from literature reviews [74] and [80] for a range of steel cutting operation, and the measurements of the shear plane angle. In the literatures, the results of the cutting tests were reported for a range of tool rake angles, +5, 0, and -7 *deg*, the cutting speed values of 100-400 *m/min* and feed values of 0.1-0.5 *mm/rev*. In this regard, prior of the shear plane angle, ϕ_c, was established from the direct measurement through the observation by a high speed camera. Additionally, prior of average friction angle, β_a, was determined, using Eq. (3.13) and the information of the literatures. Consequently, the prior mean and one standard deviation of the parameters are given as follows:

1. $\phi_c = 17 \pm 4$ *deg*
2. $\beta_a = 30 \pm 5$ *deg*
3. $\tau_s = 550 \pm 80$ *MPa*

Figure 3.10 shows the Gaussian prior distribution of ϕ_c, and Figure 3.11 illustrates the joint Gaussian prior distribution of β_a and τ_s, with the independent covariance matrix.

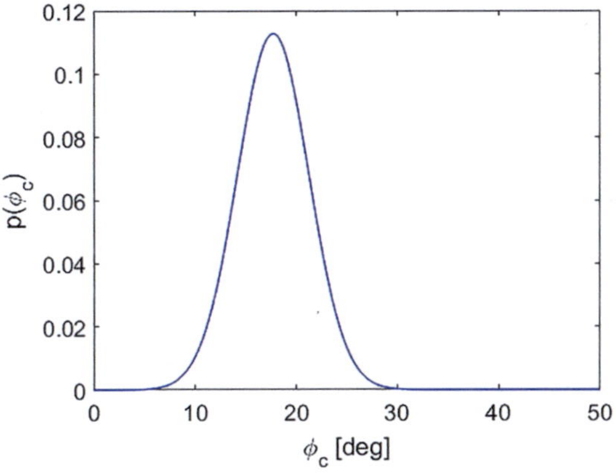

Figure 3.10: Prior distribution of ϕ_c,

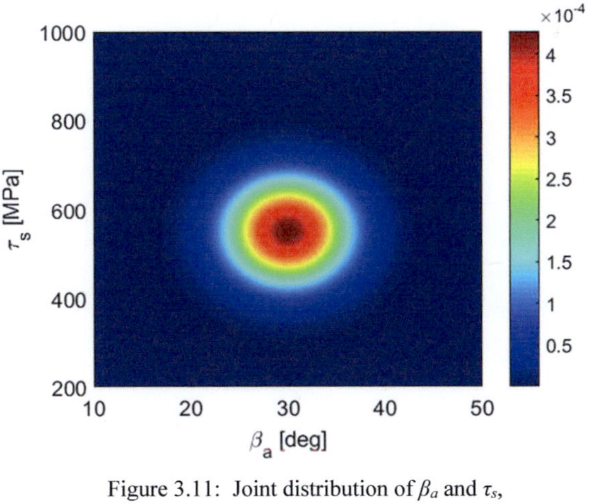

Figure 3.11: Joint distribution of β_a and τ_s,

Monte Carlo sampling was completed to find the distribution of tangential and feed force coefficients, K_t and K_f. In this case, $N = 10{,}000$ samples were drawn from the distributions of ϕ_c, β_a, and τ_s.

Figure 3.12 depicts the prior distributions of K_t and K_f, where the mean values are 2447 and 1427 *MPa* and the standard deviations are 528 and 456 *MPa*, respectively. The large uncertainty of the coefficients can be attributed to the low level of confidence in the prior of the parameters.

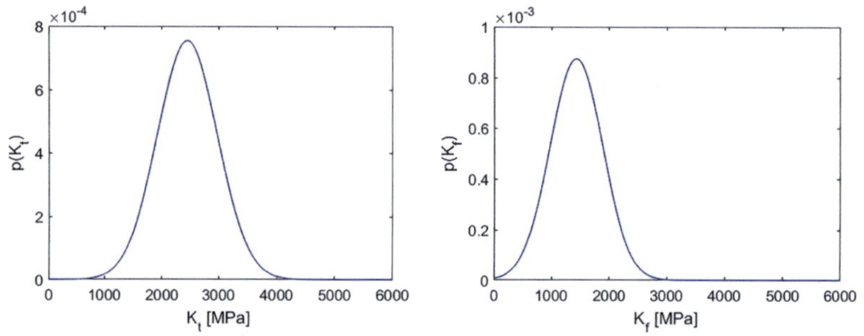

Figure 3.12: Prior distributions of K_t (left), and K_f, (right)

Once again, Monte Carlo simulation was used to represent the prior for the tangential and feed cutting forces using Eqs. (3.1) and (3.2). Figure 3.13 illustrates the functional form of the prior mean value, two standard deviations (2σ) uncertainty intervals, and the training force data points. According to the figure, the prior mean function under-estimates the forces.

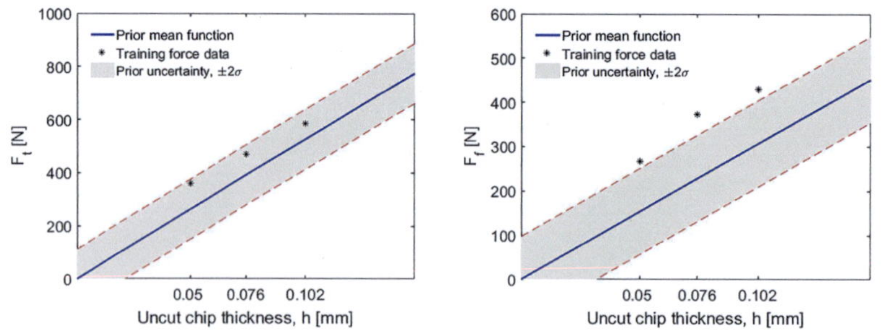

Figure 3.13: Prior functions of the tangential forces (left) and feed force (right) with $\pm 2\sigma$ standard deviations uncertainty intervals

3.5.2 Parameters Updating

This section describes the model parameters updating using the likelihood function and Metropolis MCMC method. In this regard, first, the parameter ϕ_c is updated using the measured h_c. Next, random samples from the posterior of ϕ_c, together with the measured force values are used to update the joint PDF of the parameters β_a and τ_s [74]. Likelihood function of the shear plane angle is written as follows,

$$p(\phi_c^m|\phi_c) = e^{-\frac{(\phi_c-\phi_c^m)^2}{2\sigma_{c,m}{}^2}} \tag{3.14}$$

where $\phi_c{}^m$ is the measured shear plane angle, which is calculated using measured cut chip thickness, h_c, as an input into Eq. (3.9), $\sigma_{c,m}$ is the standard deviation or variation of the measured cut chip thickness, which is obtained to be 7-10% of the thickness of measured mean value. The likelihood is the value of the PDF for the measured shear plane angle, $\phi_c{}^m$, given the specified values of the ϕ_c as prior probability. This likelihood function describes how likely the measurement result at a feed is, given the model parameters priors. In other word, if the priors result in a force, which is near to the measured force, the likelihood is high; otherwise, it is low.

Posterior distribution of ϕ_c was calculated by multiplying the prior into the likelihood function using the Metropolis algorithm. In this context, $N = 10{,}000$ samples were drawn from the proposal normal distribution, $q(\phi_c)$. After removing the first 1500 points as the burn-in period, the acceptance rate of 28% was obtained. Figure 3.14 shows posterior distributions of ϕ_c after three updates. The updated posterior mean values of the ϕ_c, in tangential and feed directions, are 10.9 and 10.8 *deg*, and the corresponding standard deviations are 0.28 and 0.29 *deg*, respectively. Accordingly, the uncertainties of the parameters are minimized.

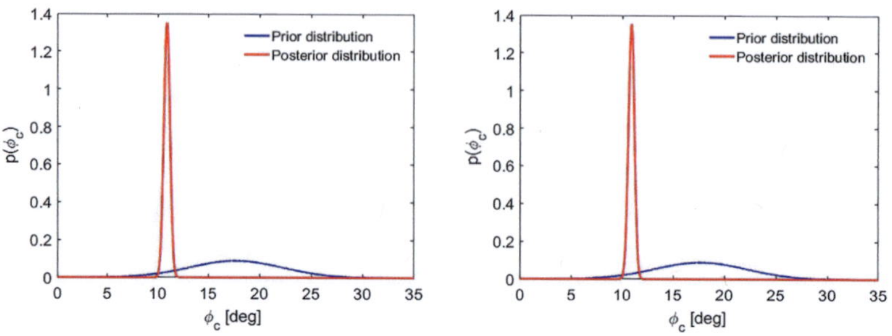

Figure 3.14: Comparison of prior and posterior distributions of ϕ_c after three updates, in tangential (left) and feed (right) directions

The same procedure was followed to update the prior joint distribution of (β_a, τ_s) using the measured force values and the samples from the ϕ_c posterior. Similarly, blockwise MCMC method was exercised to draw N = 10,000 samples from the joint normal proposal distribution, $q(\beta_a, \tau_s)$; see Algorithm 3.1. After discarding of the first 1500 samples as the burn-in period, the acceptance rate of 41% was obtained. Figure 3.15 shows the joint posterior distributions of β_a and τ_s after three updates using tangential and feed forces. The mean values of β_a and τ_s, in the tangential direction, are 30.8 *deg* and 559 *MPa*, and the standard deviations are 3.8 *deg* and 17*MPa*, respectively. The mean values of β_a and τ_s, in the feed direction, were calculated to be 34.5 *deg* and 620 *MPa*, and the standard deviations were obtained, 1.3 *deg* and 36 *MPa*, respectively. Comparing the posterior and prior joint distributions, it is shown that the MCMC simulation can reduce the uncertainty after parameters training processes.

Although the prior joint PDF of (β_a, τ_s) were taken to be independent, the parameters become correlated after running the MCMC simulation. This can be quantified using Pearson correlation coefficient. The correlation coefficient is the measure of linear relationship between two parameters defined as the covariance of the parameters divided by the product of their standard deviations; see Eq. (3.15).

$$\rho(\beta_a, \tau_s) = \frac{cov(\beta_a, \tau_s)}{\sigma_{\beta_a} \sigma_{\tau_s}} \tag{3.15}$$

The correlation coefficient of the parameters, β_a, and τ_s, in tangential and feed directions were calculated to be -0.2 and -0.67, respectively. After updating the parameters, ϕ_c, β_a, and τ_s, Monte Carlo simulation is used to calculate the posterior distribution of the coefficients, K_t and K_f, using Eq. (3.3) and (3.4).

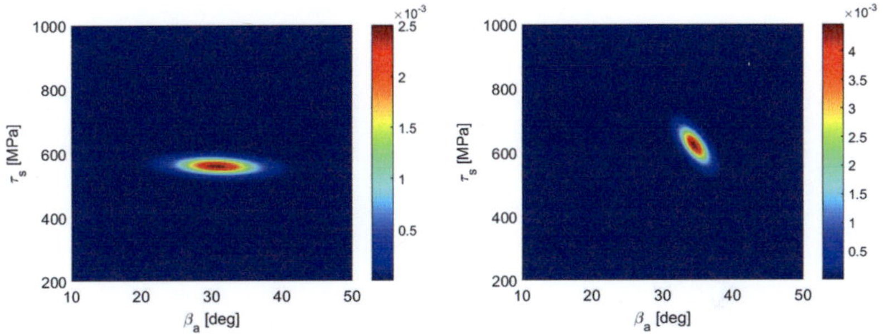

Figure 3.15: Joint PDF of β_a and τ_s after three updates using tangential (left) and feed (right) forces

Figure 3.16 shows posterior distributions of K_t and K_f, where the mean values were computed to be 3408 and 2654 *MPa*, and, the uncertainties were minimized to be 135 and 136 *MPa*, respectively.

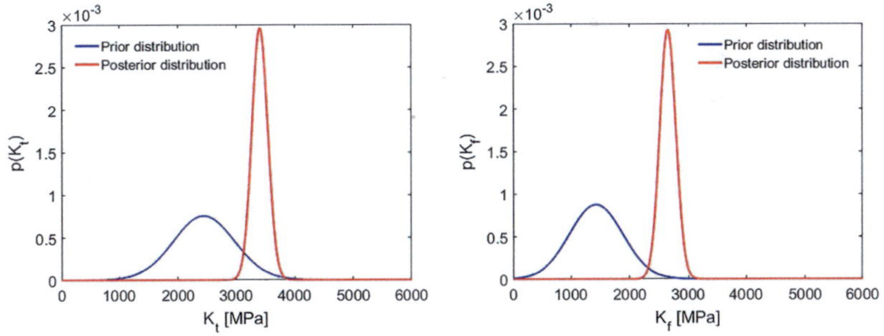

Figure 3.16: Comparison of K_t (left), *and* K_f (right), prior and posterior distributions after three updates

3.5.3 Cutting Force Prediction

Posterior forces prediction is performed inserting the posterior distributions of K_t and K_f, to the Eqs. (3.1) and (3.2). Figure 3.17 shows the functional form of the posterior tangential and feed forces with the mean and 2σ standard deviations. The uncertainties for the tangential and feed forces were quantified numerically (using the Monte Carlo method) and illustrated using the feed value of 0.076 *mm/rev*. As can be seen, despite the uncertainty intervals assignments to the posterior mean functions, the models cannot predict all of the training forces.

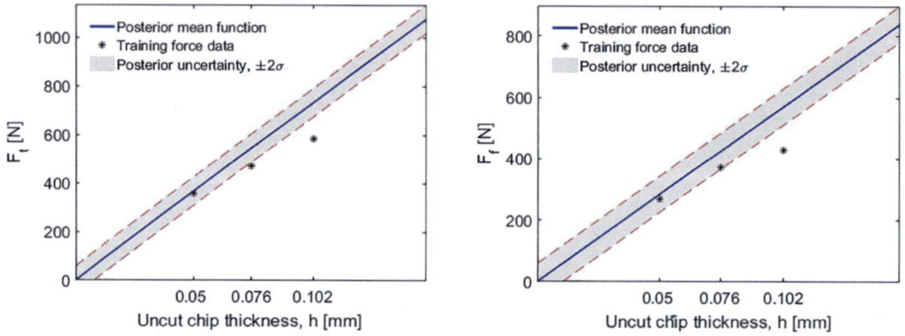

Figure 3.17: Posterior function of tangential (left) and feed (right) forces with $\pm 2\sigma$ standard deviations uncertainty intervals using Merchant model

Figure 3.18 illustrates the prediction of the cutting forces obtained under other cutting conditions using the posterior function and the corresponding credible intervals. As illustrated, the tangential and feed posterior functions can predict only a few forces.

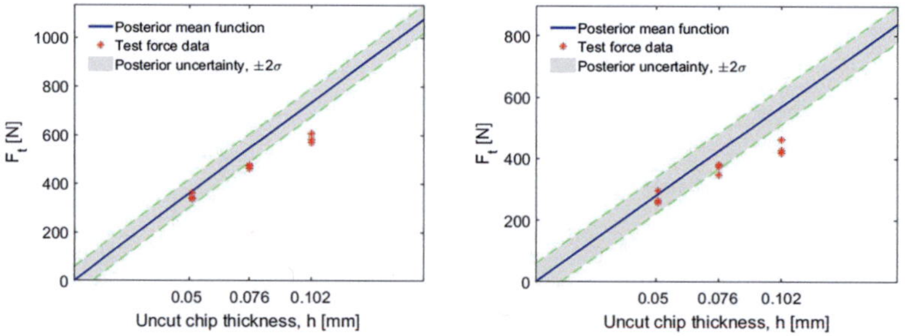

Figure 3.18: Posterior function for prediction of tangential (left) and feed (right) forces with $\pm 2\sigma$ standard deviations uncertainty intervals using Merchant model

Table 3.2 shows the cutting conditions and forces used for the prediction purpose. According to the table, each row contains one to three force values as a result of repeated tests. The reason of the imprecise prediction is that Merchant force model is based on the assumption that cutting forces F_t and F_f, are linearly proportional to the uncut chip thickness values, h. The nonlinear relationship of the forces and uncut chip thickness often appears at the low feed values. The nonlinearity can be due to the increase of the specific cutting energy with the reduced uncut chip thickness or an increase of tool edge radius. In this context, the energy is expended in shearing of the chip due to the apparent more negative effective rake angle (size effect phenomenon) [81]. The size effect is often described with the Kienzle force model [77]. Additionally, the Bayesian inference is not able to predict the forces if there is imprecision in the deterministic model. Therefore, the Merchant model can be replaced with Kienzle cutting force model, which describes a nonlinear relationship between the uncut chip thickness and the cutting force values.

Table 3.2: Cutting conditions and forces for prediction using tool rake angles 0 *deg*

No.	α_r *(deg)*	V_c *(m/min)*	*f (mm/rev)*	$F_{t\_measured}$ *(N)*	$F_{f\_measured}$ *(N)*
1	0	60	0.051		259
2	0	80	0.051	336, 341,361	263
3	0	100	0.051		297
4	0	60	0.076	472, 475	348
5	0	100	0.076	462	376, 382
6	0	60	0.102	583	419, 426
7	0	80	0.102	605	
8	0	100	0.102	567	463

3.6 Application of MCMC to Kienzle Force Model using Tool Rake Angle 0 *deg*

Blockwise Metropolis algorithm is again used to evaluate the uncertainty of the Kienzle force model parameters and the forces prediction. The uncertainty of the tangential and feed forces, F_t and F_f, originates from the uncertainty in the model parameters K_{tt}, K_{ff}, c_t, and c_f.

3.6.1 Establishing the Prior Distributions

The parameter identification starts with establishing prior values for K_{tt}, K_{ff}, c_t, and c_f. The mean and standard deviation of the parameters were taken from [82] for a range of low carbon steel cutting operations:

1. K_{tt}= 1620 ± 96 *MPa*
2. K_{ff}= 350 ± 140 *MPa*
3. c_t=0.28 ± 0.04
4. c_f=0.33 ± 0.025

Figure 3.19 shows the joint Gaussian prior distribution of K_{tt} and c_{tt}, in addition to K_{ff} and c_f, with the independent covariance matrices.

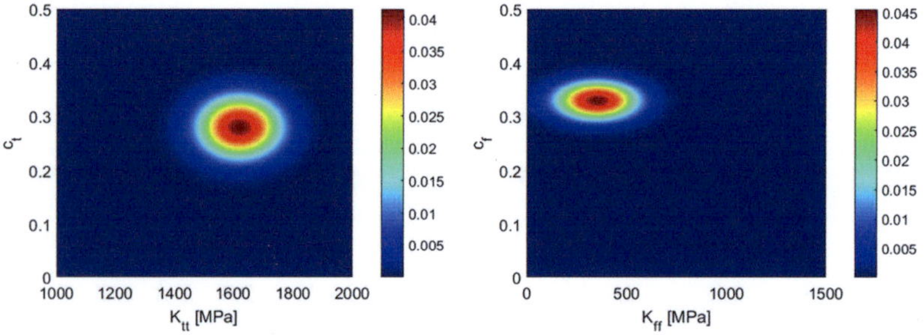

Figure 3.19: Joint prior distribution of K_{tt} and c_{tt} (left), and K_{ff} and c_f(right), for tool rake angle 0 *deg*

Table 3.3 shows the calculated mean and prior values of the tangential force, F_t, for ten different combinations of the parameters, (K_{tt},c_t). In this regard, the mean value of the force was calculated using Eq. (3.10) and the probability of each pair of the parameters (K_{tt},c_t) was computed considering the mean force value of 534 *N* and standard deviation of 70 *N* at the feed 0.076 *mm/rev*.

Moreover, the functional form of the priors mean values, two standard deviations (2σ) uncertainty intervals, and the training force data points are displayed in Figure 3.20. The training data is also shown. According to the figures, the prior mean function of the tangential force over-estimates the training force data; while the prior mean function of the feed force under-estimates them.

Table 3.3: Prior probabilities of F_t for the joint samples (K_{tt}, c_t) at feed 0.076 *mm/rev*

No.	(K_{tt}, c_t)	F_t *(N)*	*Prior probability*
1	(1530, 0.25)	466	0.0045
2	(1560, 0.25)	475	0.0049
3	(1590, 0.25)	485	0.0052
4	(1630, 0.25)	497	0.0055
5	(1680, 0.25)	512	0.0057
6	(1620, 0.26)	525	0.0056
7	(1620, 0.27)	520	0.0056
8	(1620, 0.28)	534	0.0054
9	(1620, 0.29)	547	0.0050
10	(1620, 0.3)	562	0.0044

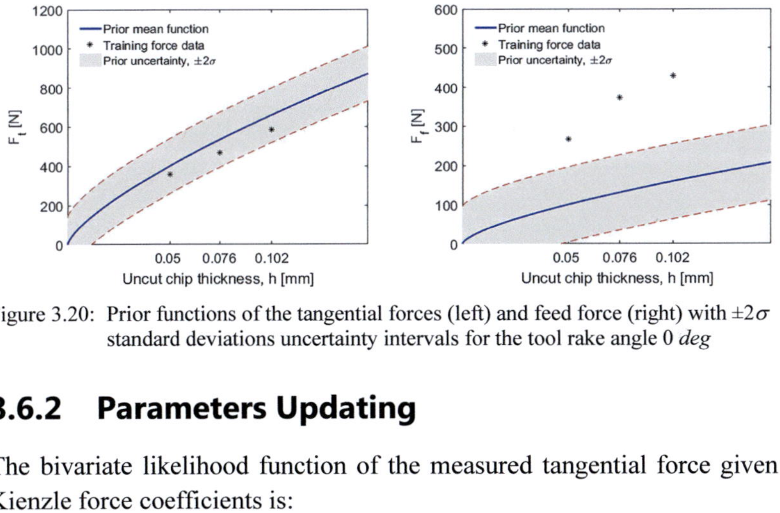

Figure 3.20: Prior functions of the tangential forces (left) and feed force (right) with $\pm 2\sigma$ standard deviations uncertainty intervals for the tool rake angle 0 *deg*

3.6.2 Parameters Updating

The bivariate likelihood function of the measured tangential force given the Kienzle force coefficients is:

$$p(F_t|\, K_{tt}, c_t) = e^{-\frac{((K_{tt}.b.f^{1-c_t})-F_t)^2}{2\sigma_{F_t}{}^2}} \tag{3.16}$$

where $p(F_t|\, K_{tt}, c_t)$ is the likelihood function of the measured mean cutting force, F_t, given specified prior values of the model coefficients, (K_{tt}, c_t), at an experimental feed value. The likelihood function is expressed as a non-normalized normal distribution, where σ_{Ft} is the standard deviation of the measured force.

To illustrate, again consider ten possible (K_{tt}, c_t) pairs listed in Table 3.4. Assume an experimental cutting force of 470 N was obtained at the feed of f = 0.076 *mm/rev*. The likelihood function can be interpreted as assigning weights to the sample forces coefficients (K_{tt}, c_t), from zero to unity, where zero means that the selected combination is not likely at all and unity means the most likely combination [65]. The likelihood for each sample force coefficients, (K_{tt}, c_t), was calculated using (3.16) considering the measured cutting force of 470 N. The value of σ_{Ft} is selected by the user's belief based on experimental force uncertainty. For this study, the standard deviation was decided to be 4-6% of the measured mean value. Table 3.4 lists the likelihood values for each possible (K_{tt}, c_t) pair. The likelihood values listed in the table imply that sample number 1 is most likely to be the correct (K_{tt}, c_t) combination, whereas sample number 10 is the least likely.

Table 3.4: Likelihood probabilities of F_t for joint samples (K_{tt}, c_t) pairs given the measured force of 468 N

No.	*(K_{tt}, c_t)*	*F_t (N)*	*Prior probability*	*Likelihood*
1	(1530, 0.25)	466	0.0045	0.0733
2	(1560, 0.25)	475	0.0049	0.0722
3	(1590, 0.25)	485	0.0052	0.0645
4	(1630, 0.25)	497	0.0055	0.0476
5	(1680, 0.25)	512	0.0057	0.0255
6	(1620, 0.26)	525	0.0056	0.0328
7	(1620, 0.27)	520	0.0056	0.0166
8	(1620, 0.28)	534	0.0054	0.0067
9	(1620, 0.29)	547	0.0050	0.0021
10	(1620, 0.3)	562	0.0044	0.0005

Figure 3.21 shows the bivariate likelihood function for F_t = 470 N at feed 0.076 *mm/rev*, given different values of (K_{tt}, c_t) pairs. The figure also demonstrates the likelihood of the joint samples number 1, which is the most likely parameters combination within the selected samples.

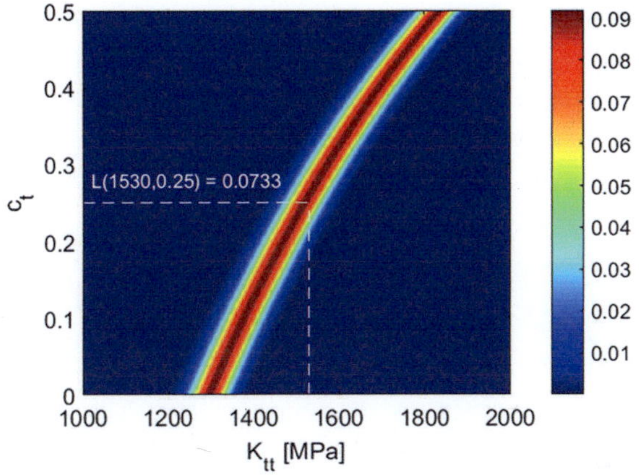

Figure 3.21: Bivariate likelihood function of the measured force 470 N at feed 0.076 mm/rev, given the (K_{tt}, c_t) pairs

Posterior distribution of the model parameters in tangential and feed directions are achieved using non-normalized product of priori and the likelihood function:

$$p(K_{tt}, c_t|F_t) = p(F_t|K_{tt}, c_t)\, p(K_{tt}, c_t) \tag{3.17}$$

where $p(K_{tt}, c_t|F_t)$ is the joint posterior distribution of the force coefficients given the measured force mean value, $p(K_{tt}, c_t)$ is the prior joint distributions of the force coefficients, $p(F_t|K_{tt}, c_t)$ is the likelihood function, and $p(F_t)$ is the normalizing factor. Joint posterior distributions, (K_{tt}, c_t) and (K_{ff}, c_f), were calculated by multiplying the priors into the likelihood functions using blockwise Metropolis algorithm. N = 10,000 samples were drawn from the proposal normal distributions, $q(K_{tt}, c_t)$ and $q(K_{ff}, c_f)$, and 1500 samples were considered as the burn-in period. The covariance matrices of the proposal distributions were tuned, so that the acceptance rate values of 44% and 33% were obtained for the drawn samples of the tangential and feed model parameters, respectively. Figure 3.22 displays the bivariate posterior distributions of K_{tt} and c_t (left), which is obtained after one update, in addition to K_{ff} and c_f (right) achieved after two updates using measured forces. For the tangential force component, the mean values of K_{tt} and c_t were computed to be 1573 MPa and 0.24, and the standard deviations are 84 MPa and 0.023, respectively. For the feed force component, the mean values of K_{ff} and c_f are 870 *MPa,* and 0.36 and the standard deviations are 58 *MPa* and 0.022, respectively. Comparing the posterior and prior joint distributions, it is seen that the uncertainties are reduced. Additionally, the model parameters become correlated with the correlation coefficient of -0.78 for (K_{tt}, c_t) and -0.85 for (K_{ff}, c_f) joint distributions.

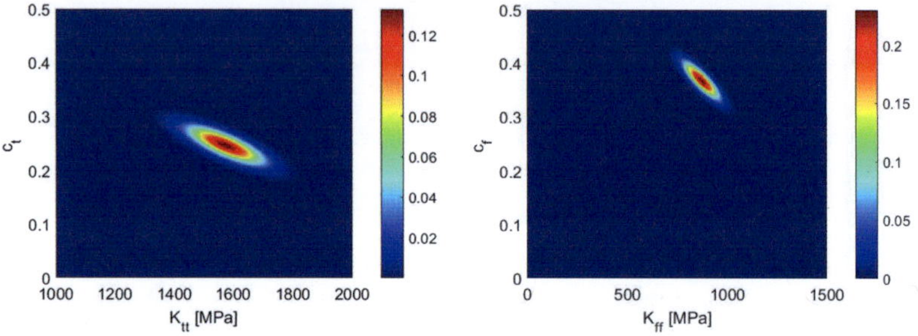

Figure 3.22: Joint posterior distribution of K_{tt} and c_{tt} (left), and K_{ff} and c_f(right), for tool rake angle 0 *deg*

3.6.3 Cutting Force Prediction

Force prediction is performed using Monte Carlo simulation and the posterior distributions of K_{tt} and c_t and K_{ff} and c_f for Eqs. (3.11) and (3.12). The uncertainty quantification is performed numerically (using the Monte Carlo method) and illustrated using the feed value of 0.076 *mm/rev.* Figure 3.23 shows the functional form of the tangential and feed forces posteriors with the mean and two standard deviations. The regression fit is characterized by $R^2 = 0.99$ (tangential force), and $R^2 = 0.98$ (feed force). As can be seen, only one force is used for updating the tangential force posterior, and two forces are used for training of the feed force posterior function. The posterior mean functions closely agree with the training forces. This is due to the influence of the informative prior knowledge in the tangential direction (which leads to the usage of only one training force) and less informative prior in the feed direction.

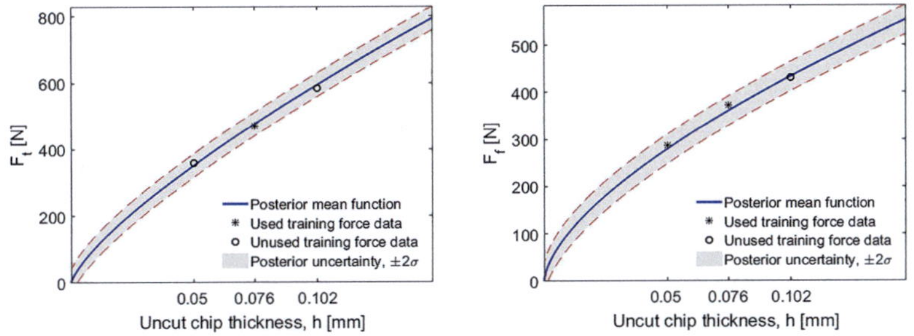

Figure 3.23: Posterior function of tangential (left) and feed (right) force with $\pm 2\sigma$ standard deviations uncertainty intervals for the tool rake angle 0 *deg*

Table 3.5 shows the prior, likelihood and posterior probabilities for each pair of (K_{tt}, c_t) pair. As can be seen, the posterior probabilities of the samples number 1 and 2 were increased compared to the prior probability. This implies that the F_t values of the 466 and 475 *N* have the highest estimation probabilities compare to the other values.

Figure 3.24 illustrates the prediction of the cutting forces obtained under other cutting conditions using the tangential and feed posterior functions. As can be seen, almost all the force data appear within the uncertainty intervals.

Table 3.5: Posterior probabilities of F_t for joint samples (K_{tt}, c_t) at feed 0.076 *mm/rev*

No.	(K_{tt}, c_t)	F_t *(N)*	*Prior probability*	*Likelihood*	*Posterior probability*
1	(1530, 0.25)	466	0.0045	0.0733	0.0134
2	(1560, 0.25)	475	0.0049	0.0722	0.0134
3	(1590, 0.25)	485	0.0052	0.0645	0.0120
4	(1630, 0.25)	497	0.0055	0.0476	0.0089
5	(1680, 0.25)	512	0.0057	0.0255	0.0049
6	(1620, 0.26)	525	0.0056	0.0328	0.0032
7	(1620, 0.27)	520	0.0056	0.0166	0.0023
8	(1620, 0.28)	534	0.0054	0.0067	0.0012
9	(1620, 0.29)	547	0.0050	0.0021	0.0004
10	(1620, 0.3)	562	0.0044	0.0005	0.0001

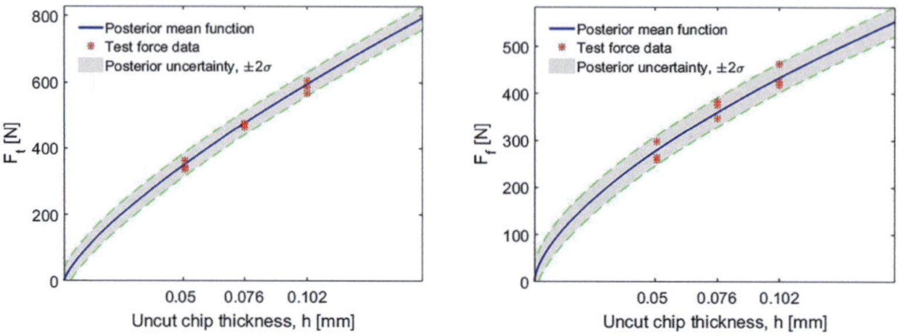

Figure 3.24: Posterior function for prediction of tangential (left) and feed (right) forces with $\pm 2\sigma$ standard deviations uncertainty intervals for the tool rake angle 0 *deg*

Table 3.6 lists the experimental force values, and the predicted mean with two standard deviations (2σ) uncertainty intervals of the tangential and feed forces for the 0 *deg* tool rake angle. Percent error values between the measured and predicted mean forces are calculated to determine the precision of the calculations described as follows,

$$\%_{error} = \left| \frac{F_{measured} - F_{predicted}}{F_{measured}} \right| \cdot 100 \tag{3.18}$$

where the maximum prediction errors for the tangential force was calculated to be 5% and for the feed force is 8%; see Table 3.6. This indicates that the algorithms are able to identify the model parameters and predict the forces with a reasonable degree of accuracy. Consequently, the probabilistic prediction of the forces using Kienzle model causes more accurate estimation and can capture the nonlinearity of the measured forces in both tangential and feed directions.

Table 3.6: Cutting conditions, measured data and predicted forces for the 0 *deg* tool rake angle

NO.	V_c (m/min)	f (mm/rev)	$F_{t\_measured}$ (N)	$F_{t\_predictedn}$ (N)	F_{t_error} (%)	$F_{f\_measured}$ (N)	$F_{f\_predicted}$ (N)	F_{f_error} (%)
1	60	0.051				259	(279, 6.7)	7.7
2	80	0.051	336, 341,361	(352, 9.67)	4.7, 3.2, 2.5	263		5
3	100	0.051				297		6
4	60	0.076	472, 475	(475 ,14.2)	0.6	348	(360, 9.9)	3.5
5	100	0.076	462		2	376, 382		4.2, 5.7
6	60	0.102	583		1.7	419, 426		3.3, 1.6
7	80	0.102	605	(593,19)	2		(433, 13.2)	
8	100	0.102	567		4.5	463		6.5

3.7 Sequential Force Prediction using Kienzle Force Model

Sequential force prediction is performed by using the posterior distributions of the Kienzle model parameters of the 0 *deg* rake tool as the prior distributions for the -10 *deg* rake tool; see Figure 3.25. As illustrated in the figure, the model parameter priors are trained by 0 *deg* rake experiments to obtain the posterior force distribution. Next, the 0 *deg* posterior distributions are used as prior probabilities for the -10 *deg* rake angle tool. The training procedure can be continued to update and predict the forces using other rake angles as well.

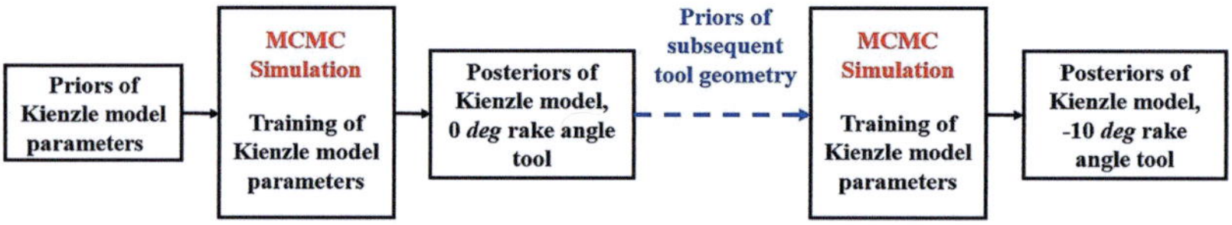

Figure 3.25: Sequential training and prediction of cutting forces using Bayesian updating for different tool rake angles

3.7.1 Model Parameters Identification

To establish the mean and standard deviation for the priors of the new geometry (tool rake angle -10 *deg*), the following steps were implemented:

1. The prior mean and standard deviation values of K_{tt} and c_{tt} parameters, for -10 *deg* rake tool, are taken to be equal to the posterior of the previous geometry.
2. The prior mean values of K_{ff} and c_f parameters, for -10 *deg* rake tool, are again taken to be equal to the posterior mean values of the previous geometry.
3. The prior standard deviations of K_{ff} and c_f parameters, for -10 *deg* rake tool, are taken to be equal to the priors of the 0 *deg* rake tool, 140 *MPa* and 0.025.

The approach mentioned above for establishing of the prior's standard deviations denotes that allocating larger uncertainty on the prior values (*i.e.*, less confidence in the prior knowledge) enables the simulation to rely more on the measurements. If more weight is given to the experiments, the parameters follow the likelihood function. On the other hand, defining smaller uncertainty on the prior distributions (more informative prior knowledge) refers that the simulation relies more on the prior. Based on this argument, it was decided to allocate smaller uncertainties to the tangential force model parameters (due to the more informative priors) and larger ones to the feed force model parameters. The functional form of the prior mean values, two standard deviations (2σ) uncertainty intervals, and the tangential and feed training force data points are shown in Figure 3.26. According to the figures, the prior mean value of the tangential force estimates better the training forces compared to the prior function of the feed force and both under-estimate the data.

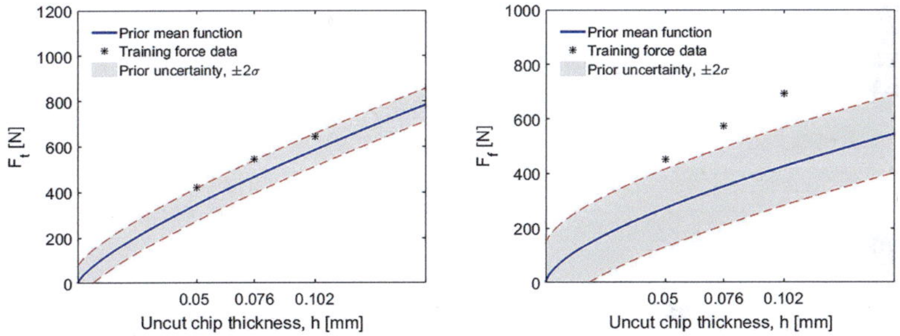

Figure 3.26: Prior functions of the tangential forces (left) and feed force (right) with $\pm 2\sigma$ standard deviations uncertainty intervals for the tool rake angle -10 *deg*

Once again, $N = 10{,}000$ samples were drawn from the proposal normal distributions, $q(K_{tt}, c_t)$ and $q(K_{ff}, c_f)$, and 1500 samples were considered as the burn-in period. The covariance matrices of the proposal distributions were tuned, so that the acceptance rate values of 45% and 39% were obtained for the drawn samples of the tangential and feed model parameters, respectively. Figure 3.27 shows the bivariate posterior distributions of K_{tt} and c_t and K_{ff} and c_f after one and two force updates, respectively. For the tangential force component, the mean values of K_{tt} and c_t were computed to be 1658 *MPa*, and 0.27 and the standard deviations are 70 *MPa*, and 0.017.

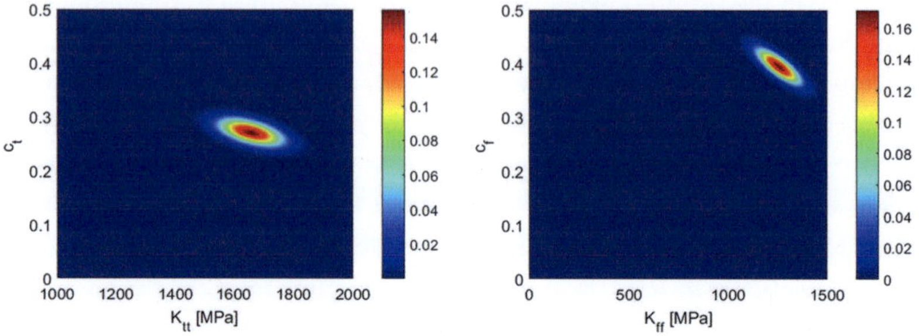

Figure 3.27: Joint posterior distribution of K_{tt} and c_{tt} (left), and K_{ff} and c_f (right) for the tool rake angle -10 *deg*

For the feed force component, the mean values of K_{ff} and c_f, are 1255 *MPa* and 0.39 and the standard deviations are 72 *MPa* and 0.021. Comparing the posterior and prior joint distributions, it is seen that the uncertainties are reduced after updating. Additionally, the correlation coefficient of the model parameters were calculated to be -0.53 for (K_{tt},c_t) and -0.80 for (K_{ff},c_f) joint distributions.

3.7.2 Cutting Forces Prediction

Posterior force prediction was performed using the posterior distributions of K_{tt} and c_t and K_{ff} and c_f for the Eqs. (3.11) and (3.12). Figure 3.28 shows the functional form of the posterior tangential and feed forces with the mean and standard deviations of 2σ. The uncertainty is quantified numerically (using the Monte Carlo method) and illustrated using the feed value of 0.076 *mm/rev*. The regression fit parameters are $R^2 = 0.96$ (tangential force), and $R^2 = 0.965$ (feed force). The posterior of the tangential force was achieved using only one update due to the more informative prior. This demonstrates the effectiveness of Bayesian inference as compared to the least squares curve fitting, which requires at least two data points for the parameter identification, in this case. On the other hand, the posterior of feed force was obtained after two updates due to its less informative prior. According to the figures, the posterior mean functions accurately represent the training forces in tangential and feed directions.

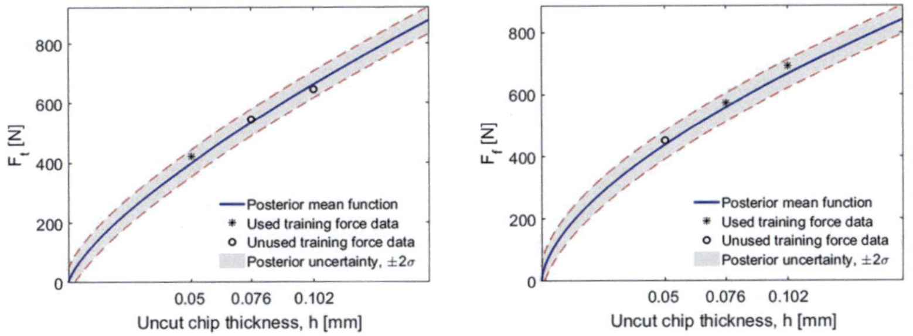

Figure 3.28: Posterior function of tangential (left) and feed (right) forces with $\pm 2\sigma$ standard deviations uncertainty intervals for the tool rake angle -10 *deg*

Figure 3.29 illustrates the prediction of the tangential and feed test forces using the posterior functions. As can be seen, all the force data appear within the uncertainty intervals.

Table 3.7 lists the experimental force values, and the predicted mean tangential and feed forces with two standard deviations (2σ) uncertainty intervals for the -10 *deg* tool rake angle. Once more, the percent error between the measured and predicted mean forces were calculated and reported in the table. The maximum prediction error for the tangential force was calculated to be 7%, and for the feed force was obtained to be 9%. This implies that the model parameters identification and forces prediction were performed with a reasonable degree of accuracy using the MCMC method applied to the Kienzle force model.

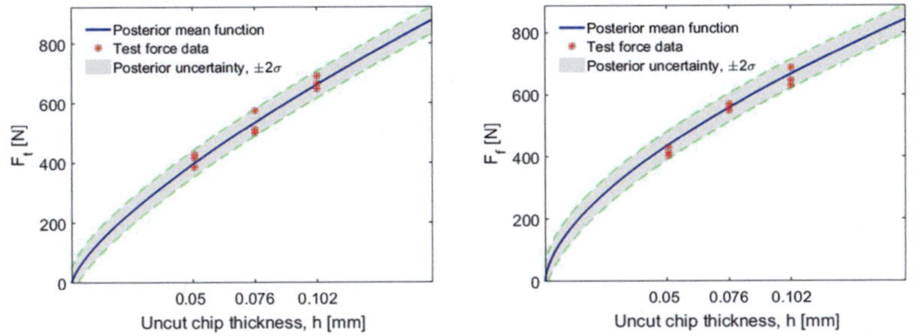

Figure 3.29: Posterior function for prediction of tangential (left) and feed (right) forces with $\pm 2\sigma$ standard deviations uncertainty intervals for the tool rake angle -10 *deg*

Table 3.7: Cutting conditions and forces for prediction for the tool rake angles -10 *deg*

N0.	V_c *(m/min)*	*f (mm/rev)*	$F_{t\_measured}$ *(N)*	$F_{t\_predictedn}$ *(N)*	F_{t_error} *(%)*	$F_{f\_measured}$ *(N)*	$F_{f\_predicted}$ *(N)*	F_{f_error} *(%)*
1	60	0.051	427		6.5	430		1.8
2	80	0.051	415	(399, 8)	3.8		(438, 8.2)	
3	100	0.051	386		3	404, 411		8.5, 6.5
4	60	0.076	576	(534, 11.8)	7.2	571	(557, 12.2)	2.5
5	80	0.076	500, 514		6.8, 3.8	549, 562		1.5, 0.8
6	80	0.102	654	(662, 15.7)	1.2	629, 646, 686	(666, 16.2)	6, 3, 3
7	100	0.102	666, 691		0.6, 4			

3.8 Conclusions

In this chapter, cutting forces prediction was performed using Bayesian inference (MCMC simulation) for the Merchant and Kienzle force models. The Mechanistic Merchant model is based on the assumption that the tool edge radius is zero, whereas the Kienzle force model takes into account the effect of cutting edge radius on specific cutting force coefficient. The results of the probabilistic force predictions using Merchant and Kienzle models for a 0 deg rake angle tool were obtained and discussed. Sequential force prediction was carried out by using the posterior probabilities of the Kienzle force model parameters for the 0 deg rake tool as the prior probabilities for the -10 *deg* rake tool. The main conclusions are summarized as follows:

1. The Kienzle force model predicted the tangential and feed cutting forces, successfully, while the Merchant model could not. The reason is that the Kienzle model can consider the size effect phenomenon in the turning process. This refers to the nonlinearity due to the increase of the specific cutting energy with the reduced uncut chip thickness or an increase of tool edge radius.
2. The Kienzle posterior functions could predict the tangential and feed forces with the good degree of accuracy for both tool geometries. Using the 0 deg rake angle tool, the maximum prediction error values were reported 5% for the tangential force and 8% for the feed force. Using the 10 deg rake angle tool, the maximum errors of 7% for the tangential force and 9% for the feed force were obtained.
3. The posterior functions of the tangential force components for both geometries were obtained using only one updating process, which is impossible in the case of parameter determination by least squares curve fitting. The uncertainty of the initial belief was reduced after updating in all instances. This suggests that Bayesian inference offers a preferred approach to force modeling by incorporating the minimal input and predicting forces under inherent uncertainties.

The result of the study can be further used to investigate the effect of cutting tools geometry and material on cutting force using Bayesian inference. The sequential probabilistic technique allows incorporating historical knowledge about process parameters into the current simulation so that the number of experiments is reduced.

4 Bayesian Updating for Cutting and Ploughing Forces Prediction in Turning Process[1]

4.1 Introduction

The semi-empirical Kienzle force model describes a nonlinear relationship between the uncut chip thickness and cutting force using a power law [83]. The model predicts the force value deterministically and, therefore, the process uncertainties, including the machining and measurement processes variability, are not inherently incorporated. They can be quantified and minimized using Bayesian inference.

Additionally, the traditional Kienzle model does not isolate the ploughing force from the cutting force. The model can be enhanced by incorporating the ploughing force component. In this chapter, orthogonal turning is performed to measure cutting forces over a range of uncut chip thickness values. An extended Kienzle force model is proposed to include the ploughing force component and nonlinear least squares fitting (LSF) method is used to identify the force model coefficients using the experimental data. Moreover, the Bayesian Markov Chain Monte Carlo (MCMC) approach is used to develop a probabilistic model. The model is verified using forces measured under other cutting conditions. Finally, the LSF and Bayesian inference predictions are compared.

---

[1] This chapter is extended from the following publication: M. Salehi, T.L. Schmitz, R. Copenhaver, R. Haas, J. Ovtcharova, Probabilistic Prediction of Cutting and Ploughing Forces using Extended Kienzle Force Model in Orthogonal Turning Process, Procedia CIRP. 77 (2018) 90-93.

4.2 Orthogonal Turning Experiments

Once more, tube turning experiments were performed on a Haas TL-1 CNC lathe; see Figure 3.2. The dry machining tests were completed using an uncoated insert SPGW09T308 with the ISO grade of *P25*, a rake angle of -10 *deg* and an edge radius of 20 *μm*. The tubular workpiece material was 1020 steel with an outer diameter of 25.4 *mm* and wall thickness of 2.1 *mm*. The corresponding chip width was 2.1 *mm*. Four feed values of f = {0.051, 0,076, 0.102, and 0.127} *mm/rev*, as well as three cutting speeds of V_c = {60, 80 and 100} *m/min*, were selected. The experiments were repeated three times for each cutting speed-feed combination. Therefore, the total number of experiments was 36. A three-axis force dynamometer (Kistler 9257B) was used to measure the cutting force. Three data sets were used to identify the force model parameters and establish the prior for the probabilistic models, while the others were used for model verification. Figure 4.1 displays the tangential force component data for identification of the force model coefficients (using non-linear LSF) and prior training purpose. The mean is provided together with one standard deviation error bars. As can be seen, the forces increase with an increase in feed.

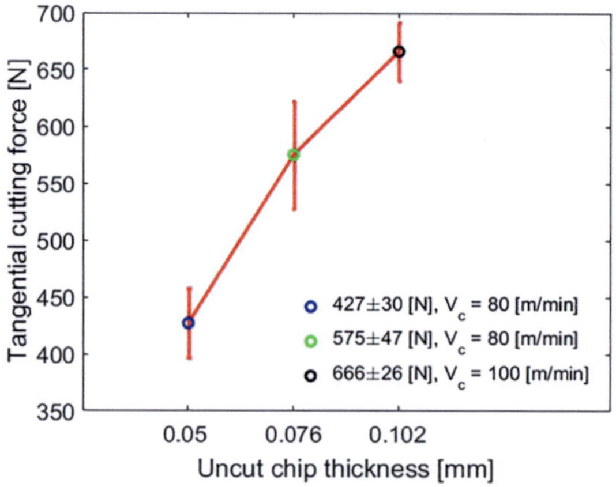

Figure 4.1: Tangential force components for training of the prior

4.3 Extended Kienzle Force Model

Kienzle force model describes a nonlinear relationship between the uncut chip thickness and the cutting force. However, since the cutting-edge corner radius is nonzero, there is an increase in chip plastic deformation without material cutting for small chip thickness values. This phenomenon is referred to as ploughing [38]; see Figure 4.2. Ploughing can be included to the force model by augmenting the Eq. (3.11), and adding a constant force coefficient that scales with the chip width, b.

$$F_t = \overbrace{K_{tt}.\, b.\, h^{1-c_t}}^{F_{t-shearing}} + \overbrace{K_{te}.\, b}^{F_{t-ploughing}} \tag{4.1}$$

In Eq. (4.1), K_{te} is associated with the tangential ploughing (rubbing) term. The shearing component is dependent on the chip thickness, while the ploughing terms is not.

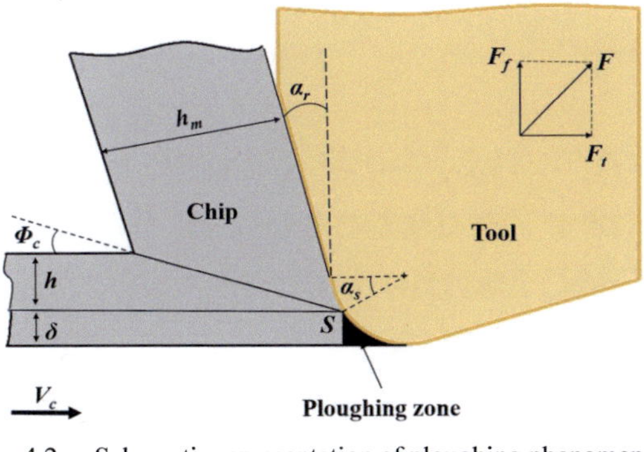

Figure 4.2: Schematic representation of ploughing phenomenon

4.4 Parameter Identification using Nonlinear LSF

The tangential force parameters, K_{tt}, K_{te}, and c_t, can be determined using the nonlinear LSF. Figure 4.3 shows the force data curve fit (using Eq. (4.1)); the regression fit quality is $R^2 = 0.987$, where the lower and upper bounds for the fit parameters were selected to be 0 and 1000. Although the fit quality is high, and three training data were used to identify the force model coefficients the approach was not able to identify the ploughing force coefficient, K_{te}.

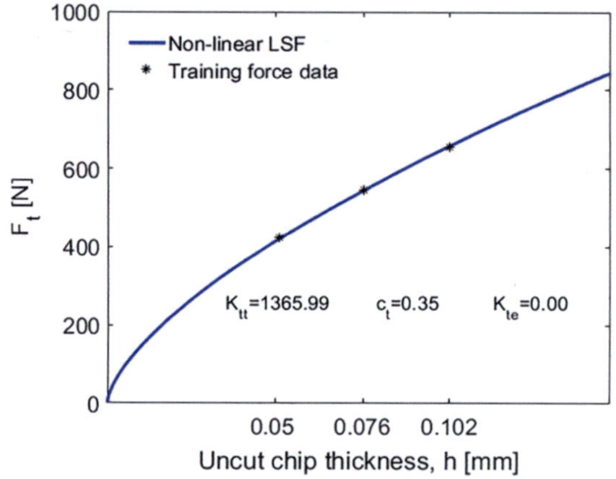

Figure 4.3: Nonlinear LSF to determine the extended Kienzle model parameters

4.5 Parameter Identification and Forces Prediction using Bayesian MCMC Method

In the extended Kienzle force model, there is uncertainty in the force value, F_t, due to the uncertainties in the model parameters, K_{tt}, K_{te}, and c_t. The uncertainties can be quantified and minimized using Blockwise MCMC method. To demonstrate, Metropolis algorithm of MCMC is presented to draw samples from the joint proposal distribution, $q(K_{tt},K_{te},c_t)$ to approximate posterior target distribution $p(K_{tt},K_{te},c_t)$; see Algorithm 4.1. According to the algorithm, the likelihood function calculates the probability of the measured tangential force, F_t, given the model parameters K_{tt}, K_{te}, and c_t. The likelihood function of the parameters, K_{tt}, K_{te}, and c_t, can be shown graphically as marginal joint PDFs of the pairs, (K_{tt},K_{te}) and (K_{tt},c_t). The posterior distribution is calculated by multiplying the prior joint distribution into the likelihood function.

Algorithm 4.1: Metropolis algorithm for updating of the extended Kienzle force model parameters

1. Establish a normal prior distribution, $p(K_{tt}, K_{te},c_t)$,
2. Establish a proposal density function for (K_{tt}, K_{te},c_t),
3. Initialize a starting sample $(K_{tt}, K_{te},c_t)^0$,
4. For i = 0 to i = N-1:

- Select a candidate $(K_{tt}, K_{te},c_t)^{new}$ from a proposal distribution, $q((K_{tt}, K_{te},c_t)^{new}| (K_{tt}, K_{te},c_t)^{i})$,
- Compute the posterior distribution, $p((K_{tt}, K_{te},c_t)^i| F_t) = p((K_{tt}, K_{te},c_t)^i)\, p(F_t \mid (K_{tt}, K_{te},c_t)^{i})$,
- Calculate the acceptance ratio, $r = \frac{p((K_{tt},K_{te},c_t)^{new})}{P((K_{tt},K_{te},c_t)^{i})}$,
- Generate a random number, $u \sim uniform\ (0,1)$,

    If $u \le r$:
        Accept the proposal: $(K_{tt}, K_{te},c_t)^{i+1} = (K_{tt}, K_{te},c_t)^{new}$,
    Else:
        Reject the proposal: $(K_{tt}, K_{te},c_t)^{i} = (K_{tt}, K_{te},c_t)^{new}$,
    End If

5. End For

4.5.1 Establishing the Prior

To develop the probabilistic model for the extended Kienzle force model, prior values for the parameters K_{tt}, K_{te}, and c_t must be selected. The mean and standard deviation of the parameters were taken from [82] for a range of low carbon steel cutting operations:

1. $K_{tt} = 1560 \pm 96\ MPa$ (one standard deviation)
2. $c_t = 0.21 \pm 0.06$ (one standard deviation)

Waldorf *et al.* [84] described the “separation point on edge” model to study the ploughing force in orthogonal cutting processes. According to the model, the separation point, *S*, of the material in front of a rounded cutting edge is defined, where the upper part converts to the chip (cut chip thickness, h_m, and shear angle, ϕ_c,) and travels along the rake face, while the lower part with the ploughing layer thickness, δ, remains attached to the workpiece; see Figure 4.2. The locating angle, α_s, for the separation point was reported to be approximately 65 *deg*. Therefore, the δ layer was calculated to be 2 μm (for a tool edge radius of 20 μm). The corresponding ploughing force coefficient can be approximated using the following steps.

a. K_t and c_t are inserted in the shearing component of Eq. (4.1) to find the cutting force,
b. The force is set equal to the ploughing component of the equation,
c. The prior value of K_{te} is found by dividing the force by the chip width, *b*.

The corresponding prior value for the ploughing force coefficient is given as follows.

3. $K_{te} = 12 \pm 2.4$ *MPa* (one standard deviation)

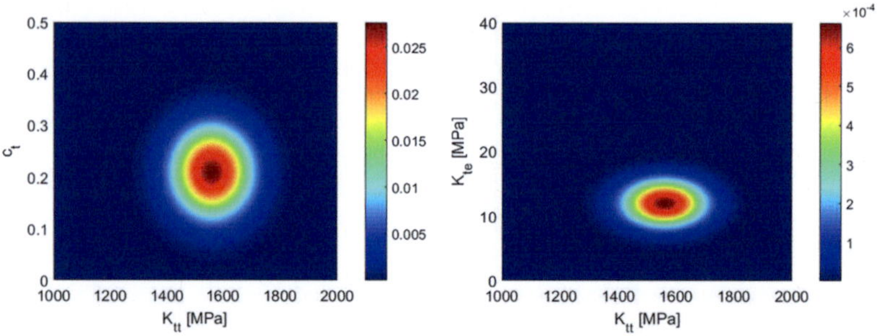

Figure 4.4: Joint prior distribution of K_{tt} and c_t (left), K_{tt} and K_{te} (right)

Figure 4.4 displays the marginal joint prior distribution of the parameters K_{tt} and K_{te} in addition to the K_{tt} and c_t. Monte Carlo simulation was used to represent the prior for the cutting force model. Figure 4.5 illustrates the functional form of the prior mean value, two standard deviations (2σ) uncertainty intervals, and the training force data points. According to the figure, the prior mean function under-predicts the forces.

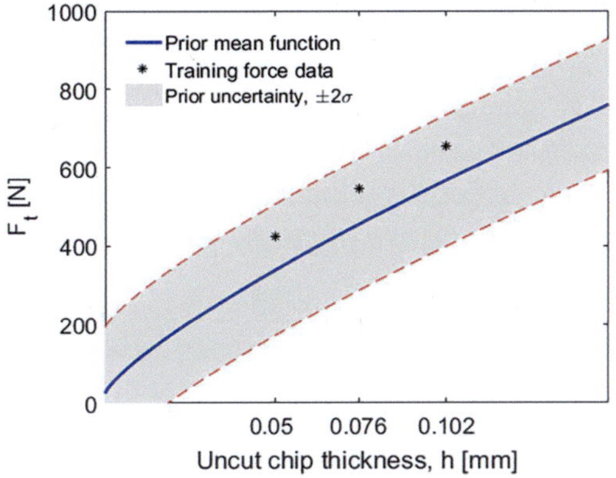

Figure 4.5: Prior function with $\pm 2\sigma$ standard deviations uncertainty intervals using extended Kienzle model

4.5.2 Likelihood Function

Figure 4.6 shows the bivariate likelihood function of the measured tangential force, F_t (427 *N*, at the feed of 0.051 *mm/rev* and the cutting speed of 60 *m/min)*, given (K_{tt}, c_t). As illustrated in the figure, the joint values with the higher probabilities are associated with the most likely values of the parameters.

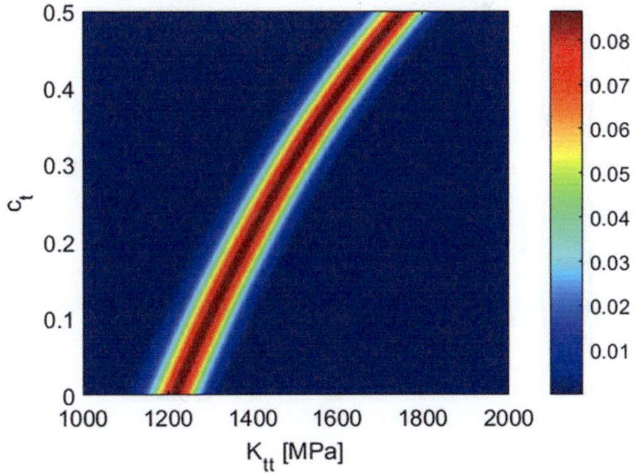

Figure 4.6: Bivariate Likelihood function of K_{tt} and c_t

Joint marginal posterior distributions, (K_{tt},c_t) and (K_{tt},K_{te}), were calculated by multiplying the priors into the likelihood functions using blockwise Metropolis algorithm. $N = 10{,}000$ samples were drawn from the proposal normal distributions, $q(K_{tt},K_{te},c_t)$, and 1500 samples were considered as the burn-in period. The covariance matrix of the proposal distribution was tuned, so that the acceptance rate of 21% was obtained for the drawn samples of the model parameter.

Figure 4.7 displays the marginal joint posterior distribution of the parameters (K_{tt}, c_t) in addition to the (K_{tt}, K_{te}). The corresponding mean values of the posterior parameters, K_t, K_{te}, and c_t, parameters, are 1597 *MPa*, 12.25 *MPa*, and 0.27, while the standard deviations are 66 *MPa*, 2.37 *MPa*, and 0.024, respectively. Moreover, the uncertainty of the posterior function is reduced. Further minimization of the uncertainties can be performed by repeating the parameters updating process using the training forces. As can be seen, the model parameters become correlated with the correlation coefficient of -0.67 for (K_{tt},c_t) and 0.01 for (K_{tt},K_{te}) joint distributions.

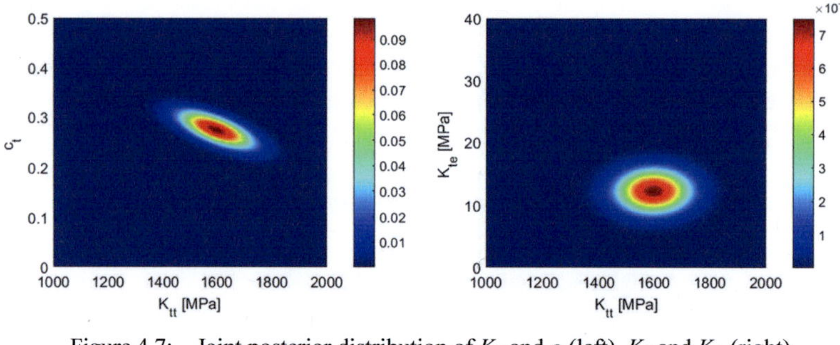

Figure 4.7: Joint posterior distribution of K_{tt} and c_t (left), K_{tt} and K_{te} (right)

4.5.3 Cutting and Ploughing Forces Prediction

Figure 4.8 depicts the functional form of the posterior distribution with the mean and standard deviations of 2σ ($R^2 = 0.989$). The uncertainty was quantified numerically (using the Monte Carlo method) and illustrated using the feed value of 0.102 *mm/rev* for the Eq. (4.1). The function approximates the shearing and ploughing components of the cutting force using only one training data, $F_t = 427\ N$. This demonstrates that Bayesian inference can identify the model parameters with the minimum input thanks to the informative prior knowledge. Nevertheless, the ploughing force identification was not successful using non-linear LSF method, despite four training force data were used. The ploughing force is shown based on extrapolation to the zero chip thickness [85]. Furthermore, the uncertainty of the posterior function is significantly reduced.

The predicted ploughing force is verified using the simplified slip-line model [86], which has been proposed for turning processes:

$$F_{tp} = \tau_s . b . r_e . \tan\left(\frac{\pi}{2} + \frac{\alpha_r}{2}\right) \tag{4.2}$$

where F_{tp} is the ploughing force, r_e is the edge radius, and α_r is the cutting edge rake angle. The τ_s parameter was computed to be 684 *MPa* using Merchant's force model [38]. From Eq. (4.2), F_{tp} is calculated to be 24.22 *N*, showing that the probabilistic model successfully approximates the ploughing force, 25.84 ± 2.37 *N*.

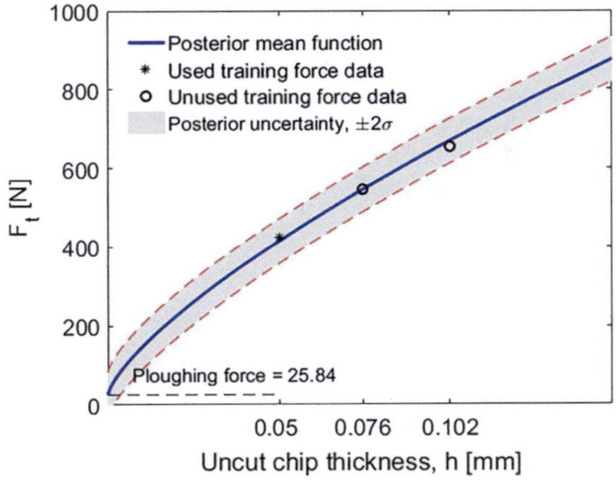

Figure 4.8: Posterior function with $\pm 2\sigma$ standard deviations uncertainty intervals using extended Kienzle model

Table 4.1 shows the cutting conditions and the measured forces in addition to the mean and one standard deviation values of the predicted forces. According to the table, each row contains one to three force values as a result of repeated tests. Percent error values between the measured and predicted mean forces are reported in the table, where the maximum error was calculated to be 7%.

Table 4.1: Cutting conditions, and measured and predicted forces

No.	V_c *(m/min)*	*f (mm/rev)*	$F_{t\_measured}$ *(N)*	$F_{t\_predicted}$ *(N)*	F_{t_error} *(%)*
1	80	0.051	415, 422	(414, 20.5)	0.2, 1.6
2	100	0.051	386		7
3	80	0.076	546, 576	(545, 24.5)	0.2, 5
4	100	0.076	500		9
5	80	0.102	647	(669, 28.3)	3.4
6	100	0.102	667, 691		0.3, 3
7	100	0.127	723, 730, 749	(779, 32)	7, 6.7, 4

Figure 4.9 illustrates the prediction of the cutting forces obtained under other cutting conditions using the posterior function. As can be seen, all the force data appear within the uncertainty intervals.

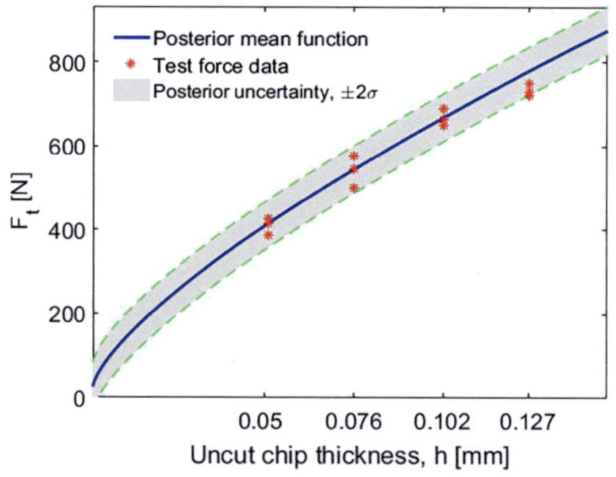

Figure 4.9: Posterior function for prediction with $\pm 2\sigma$ standard deviations uncertainty intervals using extended Kienzle model

4.6 Conclusions

Cutting and ploughing force prediction was performed using nonlinear least squares fitting and Bayesian inference (MCMC simulation) methods. The prediction results were compared with orthogonal turning data and a simplified slip-line model. Percent error values between the measured and predicted mean forces were computed, where the maximum error was obtained to be 7%. Comparing the fitting and inference approaches, it was shown that Bayesian inference could predict the cutting and ploughing forces with minimal initial data (one data point in this case) thanks to the informative prior knowledge. Furthermore, the nonlinear fitting was not able to determine the ploughing force from the cutting force despite three training force data were used and high fit quality of $R^2 \sim 0.99$ was achieved. This suggests that Bayesian inference offers a preferred approach to force modeling with minimum input and inherent uncertainty.

Part III: Bayesian-Based Analytics for Cutting Tool Life, Reliability and Wear Growth

5 Bayesian Updating for Tool Life and Reliability Analysis in Milling Process

This chapter represents the probabilistic prediction and reliability analysis of tool life applying Bayesian inference to the Taylor tool life model. The probabilistic models investigate the effect of cutting speed and tool geometry on the tool life using MCMC approach. Metropolis algorithm is used to quantify and minimize the model uncertainties. Two types of tool geometries with different edge radii were tested under a range of cutting speeds, 300-400 *m/min*, in the milling process. It is important to note that the parameters identification is valid within the mentioned range, so the tool life prediction function cannot be extrapolated beyond this range. The results of the probabilistic tool wear prediction are validated with the milling experiments, which are obtained under different cutting conditions. Tool life prediction using Bayesian and least squares curve fitting methods are compared and discussed. The tool life reliability analysis is performed using reliability and hazard functions for the range of cutting speeds and tool geometries. This chapter is organized as follows. Section 5.1 represents an introduction to the tool life modeling process and literature review. Section 5.2 reports the experimental setup and results of the milling tests. Section 5.3 presents the application of MCMC to sequential tool life prediction using the normal distribution. Section 5.4 demonstrates the reliability analysis considering the effect of cutting speeds and cutting tools geometries on the analysis. Finally, section 5.5 reports the conclusions.

5.1 Introduction

Cutting tools undergo different shapes of wear such as flank, crater, and notch wears. Flank wear is the most important and preferred form of the wear because it offers predictable and progressive tool wear pattern. This type of wear is caused due to the friction between the tool flank face and the machined surface of the workpiece [38]. Figure 5.1 shows a typical cutting tool wear pattern according to ISO 3685, where $V_{B,avg}$ denotes the average tool flank wear, $V_{B,max}$ is the maximum tool flank wear width, $V_{B,C}$ is the width of the flank wear at the tool corner, and $V_{B,N}$ is the width of notch wear. According to the ISO, the $V_{B,avg}$ exceeds 0.3 *mm*, while the $V_{B,max}$ can exceed to 0.6 *mm* [87]. Astakhov and Davim [88] has recommended the limit of the $V_{B,avg}$ to be in a range of 0.2 to 0.5 *mm* for cemented carbides in the industrial applications.

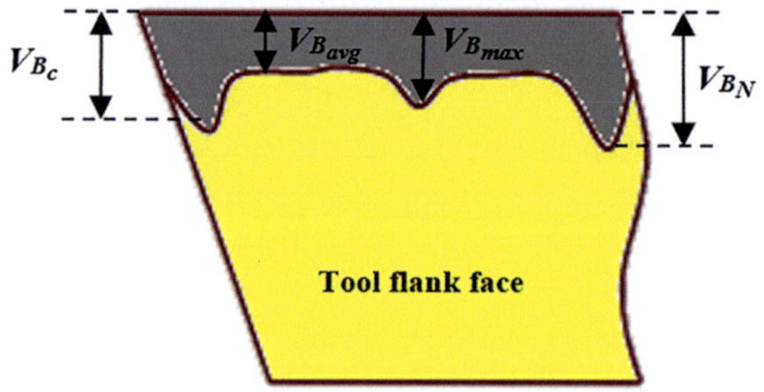

Figure 5.1: Cutting tool flank and notch wear patterns, ISO 3685

Tool life is often defined as the cutting time required to reach a predetermined wear limit of $V_{B,avg}$. Taylor equation is a useful model to predict the tool life. It describes an empirical relationship between tool life and cutting speed using the power law [65]:

$$T = \left(\frac{C}{V_c}\right)^{\frac{1}{n}} \tag{5.1}$$

where V_c is the cutting speed in *m/min*, *C* is the constant which is defined as the cutting speed required to obtain the tool life of 1 minute, *n* is the exponent that depends on the cutting parameters, and *T* is the tool life, in minute, taken to develop a certain flank wear. During the last decades, several other analytical, numerical, and empirical models have been developed to model and predict the tool wear [89–93]. However, tool wear varies in real applications, even if the cutting tool, machine, and the cutting conditions are identical. This can be due to the variation of the chemical and physical properties of the commercially identical workpiece and the tool-to-tool performance which causes to obtain stochastic quantities for same tool life tests. Therefore, the deterministic models are limited in the application if the randomness in the tool life results is not evaluated.

The uncertainty evaluation and probabilistic prediction of tool life can be performed by Bayesian inference. Karandikar *et al.* [8,69,94,95] investigated the application of grid-based and Bayesian MCMC to predict the tool life in milling and turning processes. The grid-based method was used for inference on Taylor tool life model parameters, whereas the Metropolis MCMC was applied to estimate the extended Taylor's model parameters. The performance comparison of two approaches was also reported, where the grid-based method was easier to implement, but it was computationally more expensive for updating a joint distribution with three or more dimensions. On the other hand, the Metropolis MCMC algorithm facilitated sampling from multivariate distributions without sensitivity to the number of the parameters [65]. Niaki *et al.* [67,96] developed probabilistic models using Bayesian inference to predict tool wear in the milling of Nickel-based material. The combined Gibbs-Metropolis algorithm was used to estimate the unknown parameters of a non-linear mechanistic cutting power model. The Metropolis algorithm was used for predicting the model parameters, whereas the Gibbs sampler was utilized for updating measurement error variance. By using the algorithm, the model parameters were successfully estimated, and the spindle power consumption was predicted with the maximum error of 18%.

The reliability analysis of the cutting tool life and wear have been studied using various probability distributions and statistical methods. Salonitis *et al.* [42] have used the surrogated modeling method and Stochastic Response Surface Method (SRSM) to approximate the tool flank wear in high speed turning process. Monte Carlo and First Order Reliability Methods (FORM) were used for the simulation of reliability indices. They have illustrated that the proposed technique leads to a reduction of a number of experiments in comparison to the purely empirical methods. El Wardany and Elbestawi [97] has presented stochastic models for prediction of failure rate in turning of hardened steel with ceramic tools. Different types of tool wear including, gradual wear; chemical wear and premature failure were modeled using lognormal and Weibull distributions. They investigated the failure behavior of the tool at different cutting speeds using the reliability and probability density functions, quantitatively.

Moreover, the hazard function was used to study the instantaneous failure rate, qualitatively. It was demonstrated that the lognormal distribution could be used as a tool wear distribution to predict the tool failure rate under different values of cutting conditions and workpiece surface finish. It was also shown that the Weibull distribution is the most suitable distribution to represent the tool chipping. There are some papers, which present the reliability of cutting tools under different cutting conditions and applications [98–101]. In this chapter, The Metropolis MCMC method is used for the probabilistic prediction of tool life in the milling process. Two types of tool edge radii (20 and 40 μm) are considered for the sequential prediction purpose, where the posteriors of the model's parameters of the first geometry are used as priors of the second geometry. Additionally, the effect of cutting speed and tool geometry on reliability and hazard functions of the cutting tools are investigated using the probabilistic models.

5.2 Experimental Setup, Results, and Discussion

Milling experiments were performed on a 5-axis milling machine Hermle C40; see Figure 5.2. Face milling tool wear tests (2.5 axes) were completed using tool diameter 32 *mm* holding three cutting inserts.

Figure 5.2: Tool wear measurement; milling tool and the microscope (left), and machining set up (right)

Two types of uncoated inserts (SPGW09T308) with the edge radii of 20 and 40 μm were used; see Figure 5.3. The inserts were designed and produced with the special edge geometry by Zermet Zerspanung GmbH, with the ISO grade of P25. The cubic workpiece material was AISI 1045 steel with the dimensions of 100×100×100 *mm*. The depth of cut was selected to be 1.5 *mm*. Feed value of 0.05 *mm/tooth*, and four cutting speeds of V_c = {300, 325, 350 and 400} *m/min* were selected as the cutting data.

To avoid removing the tool from the spindle, a portable microscope 5 MPX Dino-Lite (AM7915MZT) with the magnification of 220x was used to record digital images of the tool flank face in regular intervals. The tool wear was measured for three inserts after each cutting of 200 *mm* cutting length, so that the inserts wear growth was tracked.

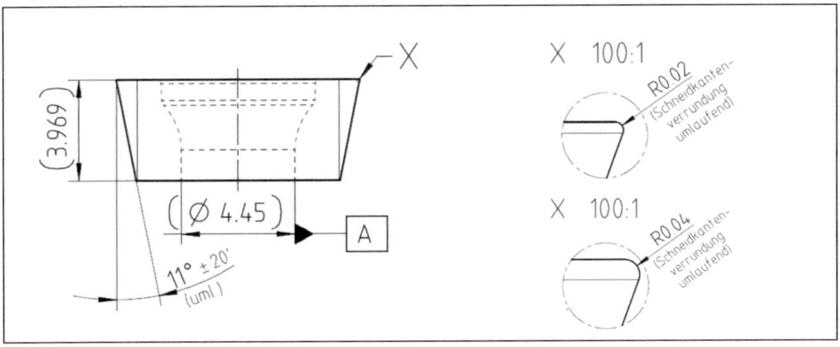

Figure 5.3: Sketches of the tool inserts, SPGW09T308 (left), and tool edge radii 20 and 40 μm (right)

To ensure the dimensional accuracy and surface quality of the specially produced inserts, the micro-geometry of the tools was inspected using a laser scanning microscope, Keyence VK8710.

Figure 5.4 shows the measurement result for the tool with a nominal edge radius of 20 μm, where, the exact radius was measured to be 19.25 μm. The measurement was repeated up to three times to calculate the corresponding mean and uncertainty. The mean and standard deviation of the edge radius were computed to be 19.27, and 0.705 μm, respectively.

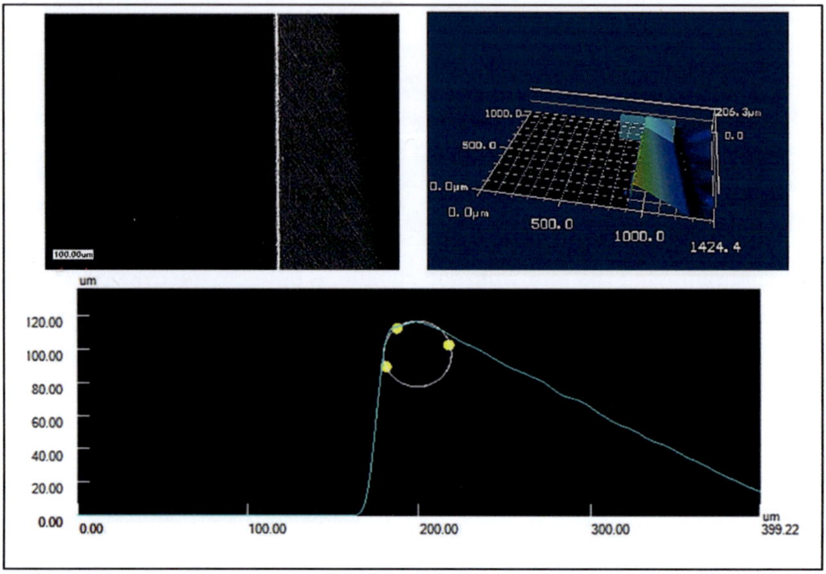

Figure 5.4: Measurement of the tool edge roundness for the tool edge radius 20 μm

Figure 5.5 displays the measurement results for the tool with a nominal edge radius of 40 μm. As illustrated in the figure, the exact radius was measured to be 42.64 μm. Again, the measurement process was repeated up to three sample tools. The mean and standard deviation of the edge radius were achieved to be 40.26, and 0.82 μm, respectively.

Cutting edge geometry is characterized by micro-geometry and edge topography. The edge topography describes the surface structure of the cutting edge. It is likely to occur microscopic damages along the cutting edge due to the grinding or sandblasting process [49,102]. The surface quality inspection is performed parallel to the cutting edge [103].

The tools edge roughness was measured along the cutting edge for three tool samples for the tool with edge radius of 20 μm, Figure 5.6. According to the results, mean and standard deviation of cutting edges based on ten-point mean roughness (Rz) were achieved to be 5.27 and 2.096 μm, respectively. The mean and standard deviation of cutting edges based on arithmetical mean roughness (Ra) were achieved to be 0.82 and 0.253 μm, respectively.

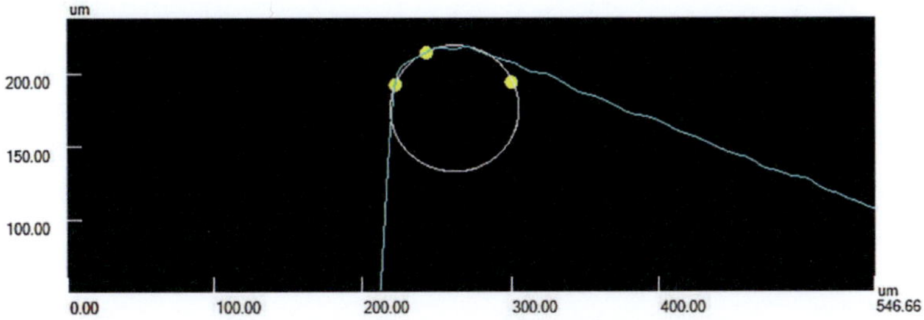

Figure 5.5: Measurement of the tool edge roundness for the tool edge radius 40 μm

The tools edge roughness was measured along the cutting edge for three tool samples for the tool with edge radius of 20 μm, Figure 5.6. According to the results, mean and standard deviation of cutting edges based on ten-point mean roughness (Rz) were achieved to be 5.27 and 2.096 μm, respectively. The mean and standard deviation of cutting edges based on arithmetical mean roughness (Ra) were achieved to be 0.82 and 0.253 μm, respectively.

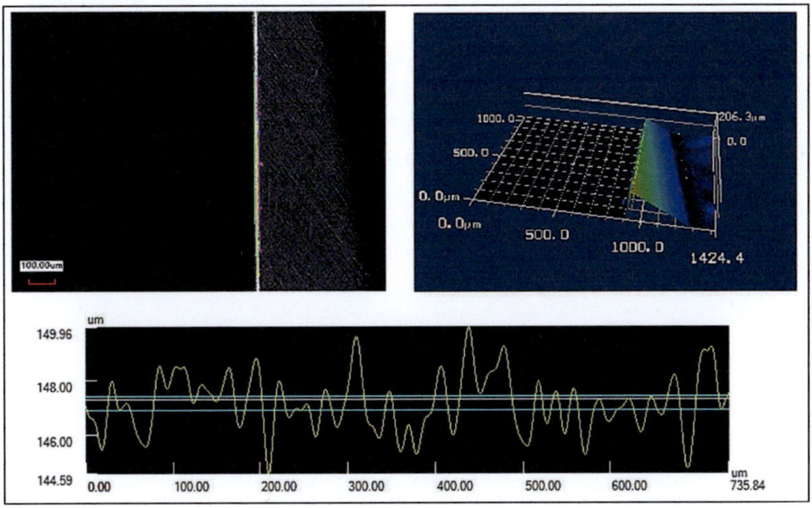

Figure 5.6: Measurement of the tool edge roughness for the tool edge radius 20 μm

Figure 5.7 displays the result of the edge roughness measurement for the tool with edge radius 40 μm. According to the results, mean and standard deviation of cutting edges based on ten-point mean roughness (R_z) were achieved to be 11.59 and 0.978 μm, respectively. Moreover, the mean and standard deviation of the cutting edges based on arithmetical mean roughness (R_a) were obtained to be 1.47 and 0.319 μm, respectively.

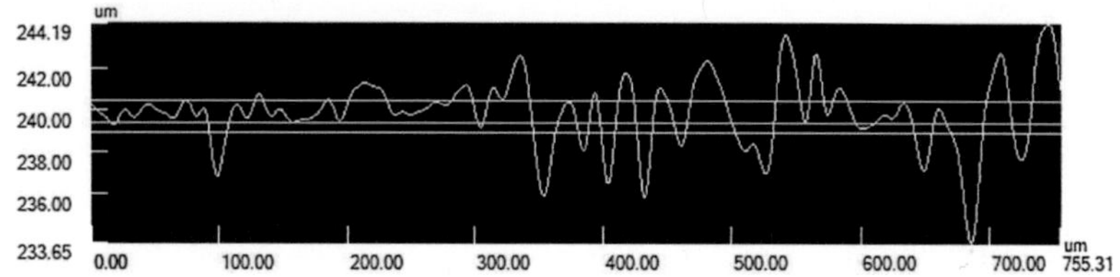

Figure 5.7: Measurement of the tool edge roughness for the tool edge radius 40 μm

5.2.1 Tool Life and Wear Growth Measurement

Average tool flank wear, $V_{B,avg}$, was measured in regular intervals until the flank wear criterion 0.3 *mm* is reached. Figure 5.8 illustrates the tool wear growth images recorded at five intervals using the tool edge radius of 20 μm, the cutting speed of 350 *m/min*, and the feed at 0.05 *mm/tooth*. The wear amount of each data point and the corresponding cutting time was recorded and illustrated. To establish the tool life for each test, linear interpolation was used between adjacent data points until the wear limit of 0.3 *mm* is reached.

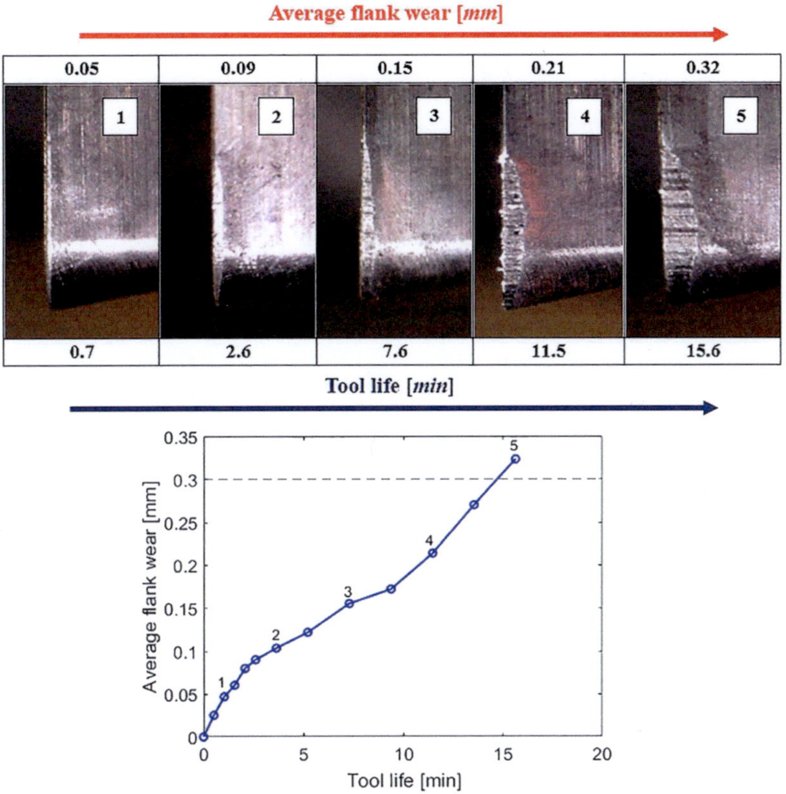

Figure 5.8: Average flank wear growth using tool edge radius 20 *μm*, cutting speed 350 *m/min*, and feed 0.05 *mm/tooth*

The results of the tool wear tests using the tool edge radii of 20 and 40 *μm* are demonstrated in Figure 5.9 and Figure 5.10, respectively. For each data point, the mean is provided together with one standard deviation error bars as a result of three measurements (three inserts wear amount) for each cutting pass. As can be seen, the tool life reduces with an increase in cutting speed for both geometries. Moreover, the tool life using the tool edge radius of 20 *μm* is longer than the edge radius of 40 *μm*. Table 5.1 shows the measured mean values of the tool life at the average flank wear of 0.3 *mm* and the feed value of 0.05 *mm/tooth*.

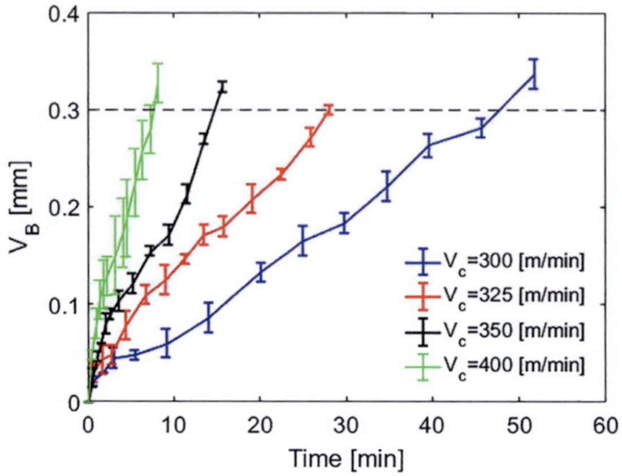

Figure 5.9: Tool flank wear growth using the tool edge radius of 20 *μm*, and feed 0.05 *mm/tooth*

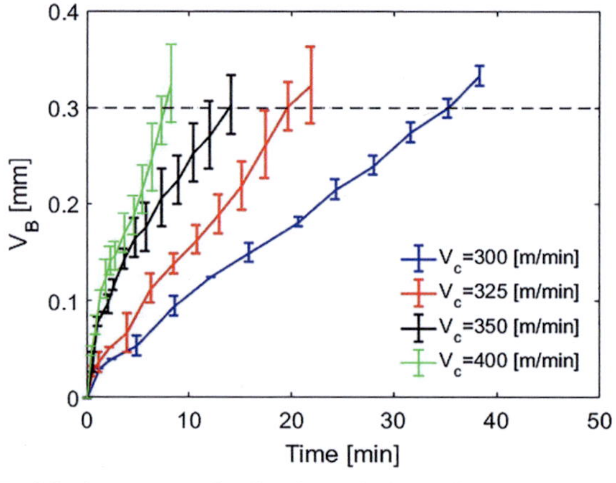

Figure 5.10: Tool flank wear growth using the tool edge radius of 40 *μm*, and feed 0.05 *mm/ tooth*

Table 5.1: Mean value of tool life at the average flank wear of 0.3 mm and the feed of 0.05 mm/ tooth

No.	*Tool edge radius (μm)*	*V_c (m/min)*	*Measure tool life (min)*
1	20	300	47.6
2	20	325	28
3	20	350	14.7
4	20	400	7.7
5	40	300	36.2
6	40	325	19.5
7	40	350	13.8
8	40	400	7.6

5.3 Probabilistic Sequential Prediction of Tool Life

Probabilistic prediction of tool life using normal distribution is performed starting with establishing priors to the Taylor tool life model parameters, C and n. In the Taylor model, the uncertainty of the tool life is due to the uncertainties in the model parameters, C and n. The priors are then updated by the result of wear test using the tool radius of 20 μm to obtain the posterior distribution of the parameters and tool life. The posteriors of the model parameters for the 20 μm edge radius tool are used as priors of the subsequent tool geometry, 40 μm edge radius tool. The results of the simulation are validated with the measured tool life under other cutting conditions. Finally, the model parameter identification using MCMC and least squares fit method are compared to each other.

5.3.1 Application of MCMC to the Tool Edge Radius 20 μm

Bayesian Markov Chain Monte Carlo is used to estimate the Taylor model parameters. Metropolis algorithm of MCMC is again used to draw samples from the joint proposal distribution $q(C,n)$ to approximate the posterior target distribution, $p(C,n)$; see Algorithm 5.1. According to the algorithm, the likelihood function calculates the probability of the measured tool life, T_m, given the model parameters, (C,n). The posterior distribution is calculated by multiplying the prior joint distribution into the likelihood function.

Algorithm 5.1: Metropolis algorithm for updating the Taylor tool life model parameters

1. Establish a normal prior distribution, $p(C,n)$,
2. Establish a proposal density function for (C,n),
3. Initialize a starting sample $(C,n)^0$,
4. For i = 0 to i = N-1:
- Select a candidate $(C,n)^{new}$ from a proposal distribution, $q((C,n)^{new} | (C,n)^{i})$,
- Compute the posterior distribution, $p((C,n)^{i} | T_m) = p((C,n)^{i})\, p(T_m | (C,n)^{i})$,
- Calculate the acceptance ratio, $r = \frac{p((C,n)^{new})}{P((C,n)^{i})}$,
- Generate a random number, $u \sim uniform\ (0,1)$,

  If $u \leq r$:

    Accept the proposal: $(C,n)^{i+1} = (C,n)^{new}$,

  Else:

    Reject the proposal: $(C,n)^{i} = (C,n)^{new}$,

  End If

5. End For

5.3.1.1 Establishing the Prior

The parameter identification starts with establishing prior values for the parameters C and n as a joint Gaussian distribution. The mean and one standard deviation of the parameters were taken from [65,82] for a range of low carbon steel cutting operations:

1. $C = 340 \pm 60$ *m/min*
2. $n = 0.26 \pm 0.05$

Figure 5.11 displays the joint priors of the pairs (C,n), where the covariance matrices were taken to be independent. Monte Carlo simulation was exercised to determine the prior distribution of the tool life, T, using the joint priors of the C and n. For the simulation, $N = 3000$ random C and n samples were drawn from the prior joint distribution for the Eq. (5.1) using the cutting speed of 325 *m/min*.

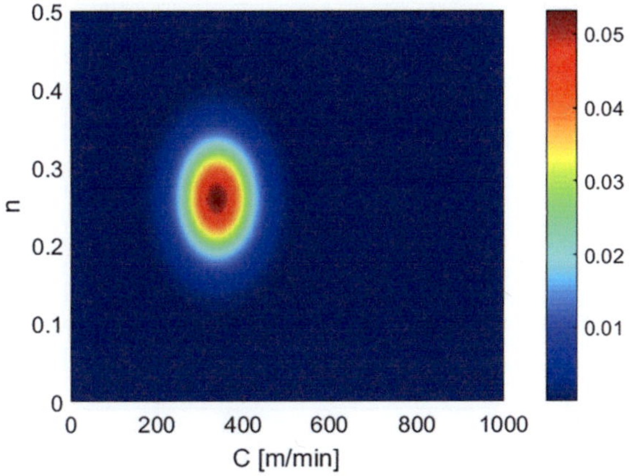

Figure 5.11: Prior joint Gaussian distribution of the parameters C and n

Figure 5.12 displays the histogram of the tool life (in blue) and a lognormal distribution fit to the histogram (solid red line), where the histogram is normalized to obtain a unit area under the curve. As shown in the figure, the lognormal distribution provides a good approximation to the histogram of the tool life prior. Note that the right-skewed distribution of the PDF is obtained due to the power law form of the Taylor tool life equation. The lognormal distribution represents the prior PDF of tool life at the cutting speed of 325 m/min.

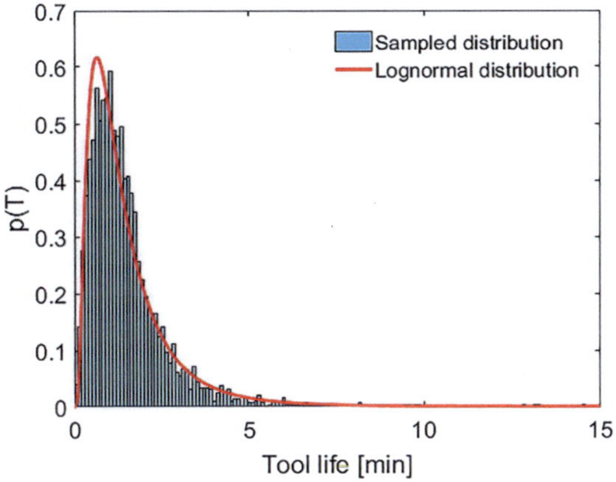

Figure 5.12: Prior distribution of tool life using the cutting speed of 325 *m/min*

The functional form of the tool life prior with the median and the confidence intervals are displayed in Figure 5.13. The lognormal distributions were used to illustrate the uncertainty intervals of tool life within the selected range cutting speeds. Since the median value is associated with the 50th percentile of the PDF, the prior function is characterized by the median and the bounds of the credible intervals. The lower and upper bounds of the credible intervals were taken to be 15.87th and 84.13th percentiles, respectively. The tool life data is also shown. According to the figure, the function underestimates the data points.

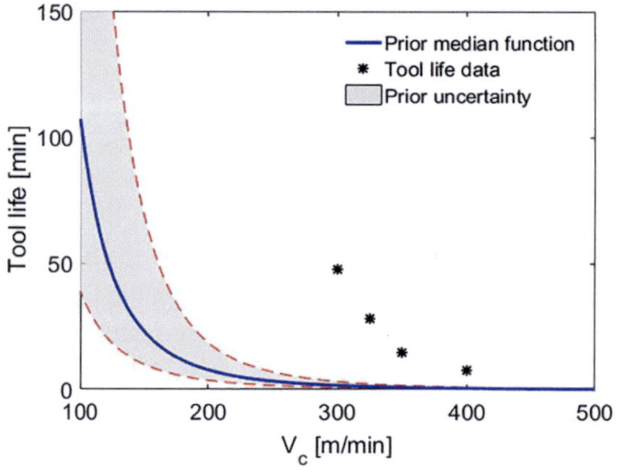

Figure 5.13: Prior function of the tool life using 20 μm edge radius tool

The prior probability of tool life can be computed for each drawn sample from the joint PDF of the parameters *C* and *n*. To demonstrate the procedure, assume the case that ten samples of *(C, n)* pairs are selected and inserted to the Eq. (5.1) to calculate the tool life values; see Table 5.2. According to table, the corresponding probabilities for the selected samples are computed using the Eq. (2.6). In this case, the tool life probability of each pair was calculated using the median tool life value of 1.2 *min* and a standard deviation of 0.63 *min* at the cutting speed of 325 *m/min*. As can be seen, the tool life value of the second sample, 0.94 *min*, has the highest probability of 0.57; whereas, the sample number 5 has the lowest probability of 0.04.

Table 5.2: Prior probabilities for the joint samples (C, n) using the cutting speed of 325 *m/min*

No.	*(C,n)*	$T_{\_predicted}$ *(min)*	*Prior probability*
1	(280, 0.26)	0.56	0.37
2	(320, 0.26)	0.94	0.57
3	(360, 0.26)	1.48	0.56
4	(400, 0.26)	2.22	0.17
5	(400, 0.21)	2.68	0.04
6	(400, 0.23)	2.46	0.09
7	(400, 0.25)	2.3	0.14
8	(400, 0.27)	2.15	0.2
9	(400, 0.29)	2.04	0.26
10	(400, 0.31)	1.95	0.31

5.3.1.2 Parameters Updating

MCMC Simulation was exercised to achieve the posterior distribution of the parameters C and n. For the simulation, $N = 10{,}000$ samples were drawn from the joint normal proposal distribution, $q(C,n)$. The samples are then updated by the results of the experiments (likelihood function); see Algorithm 5.1. Bivariate likelihood function of the measured tool life given the Taylor model parameters is written as follows:

$$p(T_m|C,n) = e^{-\frac{\left(\left(\frac{C}{V_c}\right)^{\frac{1}{n}}-T_m\right)^2}{2\sigma_{Tm}{}^2}} \tag{5.2}$$

where $p(T_m|\, C,n)$ is the likelihood function of the measured tool life, T_m, given specified prior values of the Taylor model coefficients, (C,n), at an experimental cutting speed value, V_c. The likelihood function is expressed as a non-normalized normal distribution, where σ_{Ttm} is the standard deviation of the measured tool life. This likelihood function describes how likely the measurement result at a particular cutting speed is, given prior parameters values. In other word, if priors result in a tool life, which is near to the measured tool life, the likelihood is high; otherwise, it is low.

To illustrate the functionality of the likelihood function, consider ten possible (C,n) pairs listed in Table 5.3. Assume an experimental tool life of 28 *min* was obtained at cutting speed, V_c = 325 *m/min*. The likelihood function can be interpreted as assigning weights to the sampled parameters (C,n), from zero to unity, where zero means that the selected combination is not likely at all and unity means the most likely combination. The likelihood for each sample pairs, (C,n), was calculated using Eq. (5.2) given the measured tool life of 28 *min*. The value of σ_{Tm} was selected by the user's belief based on experimental tool life uncertainty. For this study, the standard deviation was decided to be 5-7% of the mean value.

Table 5.3: Prior probabilities and likelihood for the joint samples (C, n) using cutting speed 325 *m/min*

No.	*(C,n)*	$T_{\_predicted}$ *(min)*	*Prior probability*	*Likelihood*
1	(320, 0.16)	0.9	0.56	4.65×10^{-22}
2	(360, 0.16)	1.9	0.34	1.32×10^{-19}
3	(400, 0.16)	3.66	3.83×10^{-14}	3.85×10^{-17}
4	(440, 0.16)	6.42	1×10^{-16}	2.26×10^{-13}
5	(480, 0.16)	11.4	8.24×10^{-57}	2.46×10^{-8}
6	(540, 0.16)	23.8	8.24×10^{-257}	0.33
7	(540, 0.20)	12.6	6×10^{-71}	2.95×10^{-7}
8	(540, 0.24)	8.3	9.95×10^{-28}	1.7×10^{-11}
9	(540, 0.28)	6.13	7.26×10^{-14}	5.35×10^{-14}
10	(540, 0.32)	4.88	3.7×10^{-8}	1.56×10^{-15}

Table 5.3 lists the likelihood values for each possible *(C,n)* pair. Comparing the prior and the likelihood values, it is seen that the joint samples number 5 had the lowest prior probability before the update, and now has the highest likelihood after the update. This implies that the MCMC simulation shifts the parameters C and n towards the measured tool life value of 28 *min*, so the parameters are strongly influenced by the likelihood function. On the other hand, joint samples number 1 had the highest prior probability before the update and now has the lowest likelihood after the update. This sample is rejected by the Metropolis algorithm because of the negligible influence on the posterior distribution. This causes the simulation to rely on those prior values, which are closer to the measured tool life quantities.

MCMC simulations are continued by updating the model parameters using the second measured tool life 7.7 *min*, at the cutting speed of 400 *m/min*. Figure 5.14 shows the joint posterior of the parameters *C* and *n* (right). The mean values of the parameters *C* and *n* are 543 *m/min* and 0.155, and the parameters standard deviation values are 21.5 *m/min* and 0.014, respectively. Comparing the posterior and prior joint distributions, it is seen that the uncertainties are reduced. Additionally, the model parameters become correlated with the correlation coefficient of 0.95.

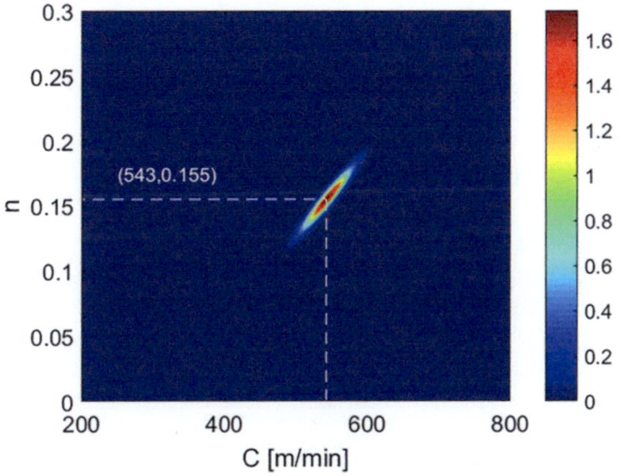

Figure 5.14: Posterior joint Gaussian distribution of the parameters *C* and *n* for the 20 μm edge radius tool

As mentioned in the section 2.4.3, convergence of the Markov chain can be improved using burn-in technique. Figure 5.15 shows the trace plot of the burn-in period for the parameters *C* and *n* with arbitrary starting points of the 300 and 0.3, respectively. To ensure the convergence of the chain, 1000 initial samples were discarded from the iterations, so that the remaining samples approach the steady state condition. As can be seen in the figures, the chain approaches to the posterior mean of the parameters *C* and *n* after the transient iterations.

Figure 5.16 displays the marginal posterior distribution of the parameter *C* along with the trace plot for the entire iterations. The figure shows that the Metropolis rejecting/accepting algorithm approved 34% of the drawn samples, which is a reasonable acceptance rate [56].

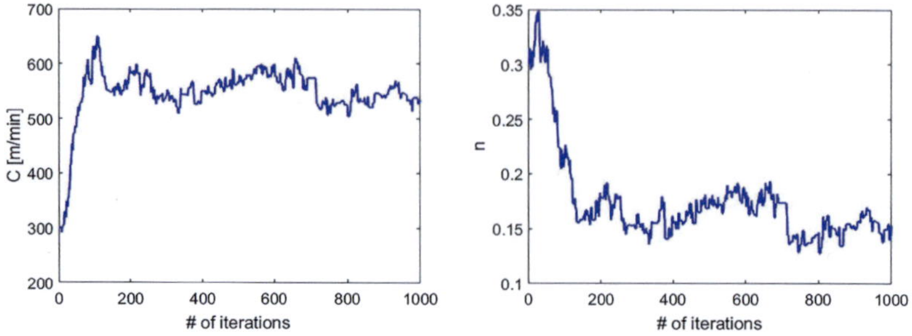

Figure 5.15: Trace plots of the burn-in period for parameters C and n using 20 μm edge radius tool

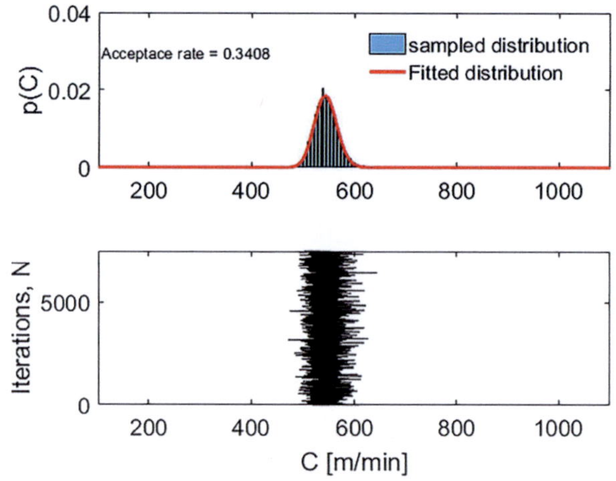

Figure 5.16: Marginal posterior distribution and trace plot of the parameter C

5.3.1.3 Tool Life Prediction

Posterior tool life prediction is performed using Monte Carlo simulation by inserting the marginal posterior distributions of C and n to the Eq. (5.1). The uncertainty quantification is performed numerically (using Monte Carlo method) for the range of cutting speeds, 275-425 *m/min*. Figure 5.17 demonstrates the functional form of the posterior tool life with the mean and standard deviations of 2σ and the regression fit quality of $R^2 = 0.991$. It important to note that prior lognormal distribution is converted to the posterior normal distribution because of the influence of Gaussian likelihood function (*i.e.*, central limit theorem); see section 2.4.1. As can be seen, the posterior mean functions approximate the training (black points) and tests data (red points), accurately. It is seen that the posterior function is achieved using two training points. It is important to note that the posterior function can only predict the data points within the predetermined cutting speed range 300-400 *m/min* and extrapolation before and after the range is not valid.

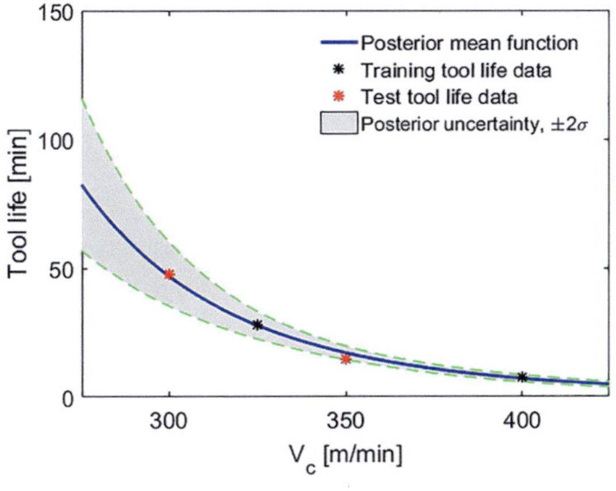

Figure 5.17: Posterior function with $\pm 2\sigma$ standard deviations uncertainty intervals using 20 μm edge radius tool

Table 5.4 shows the prior, likelihood and posterior probabilities for each *(C,n)* pair. As can be seen, the posterior probability of the joint samples number 6 is maximized from the prior probability of 8.24×10^{-257} to the 0.0575, thanks to the influence of the larger likelihood value. On the other hand, the minimum posterior probability belongs to the joint samples number 1 due to the weak influence of the likelihood function. This implies that, in this case, the posterior probability is strongly influenced by likelihood function (experiments results) due to the less informative priors.

Table 5.4: Prior, likelihood and posterior probabilities for (*C, n*) using cutting speed of 325 *m/min*

No.	*(C,n)*	$T_{\_predicted}$ *(min)*	*Prior probability*	*Likelihood*	*Posterior probability*
1	(320, 0.16)	0.9	0.56	4.65×10^{-22}	1.44×10^{-7}
2	(360, 0.16)	1.9	0.34	1.32×10^{-19}	3.72×10^{-7}
3	(400, 0.16)	3.66	3.83×10^{-14}	3.85×10^{-17}	1.86×10^{-6}
4	(440, 0.16)	6.42	1×10^{-16}	2.26×10^{-13}	2.17×10^{-5}
5	(480, 0.16)	11.4	8.24×10^{-57}	2.46×10^{-8}	5.97×10^{-4}
6	(540, 0.16)	23.8	8.24×10^{-257}	0.33	0.0575
7	(540, 0.20)	12.6	6×10^{-71}	2.95×10^{-7}	0.0012
8	(540, 0.24)	8.3	9.95×10^{-28}	1.7×10^{-11}	7.41×10^{-5}
9	(540, 0.28)	6.13	7.26×10^{-14}	5.35×10^{-14}	1.46×10^{-5}
10	(540, 0.32)	4.88	3.7×10^{-8}	1.56×10^{-15}	5.32×10^{-6}

Figure 5.18 displays the comparison of lognormal prior and normal posterior PDFs and tool life data points at the average tool wear of 0.3 *mm*. As illustrated in the figure, the posterior distribution is able to predict the tool life data at the cutting speed of 300 *m/min*, accurately.

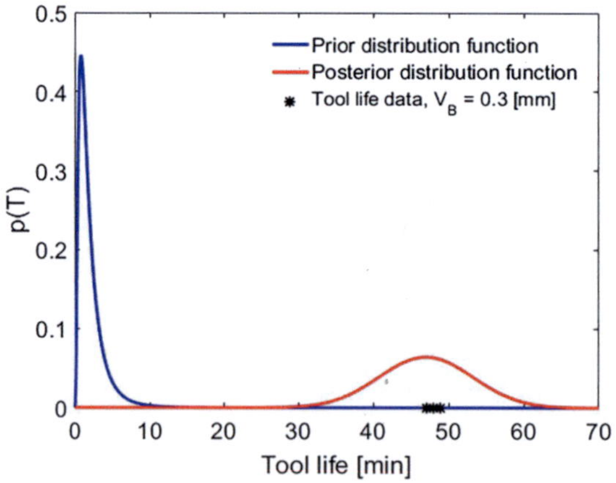

Figure 5.18: Prior and posterior PDFs using cutting speed 300 *m/min* using 20 *μm* edge radius tool

5.3.2 Application of MCMC to the Tool Edge Radius 40 μm

Sequential probabilistic tool life prediction is performed by using the posterior distributions of the Taylor model parameters, C and n, (from the 20 μm edge radius tool) as priors for the 40 μm edge radius tool; see Figure 5.19. As illustrated in the figure, the model priors are trained by20 μm edge radius tool experiments to obtain the tool life posterior distributions, first. Second, the 20 μm posterior distributions are used as priors for the 40 μm edge radius tool. The training procedure can be continued to update and predict the forces using further edge radii.

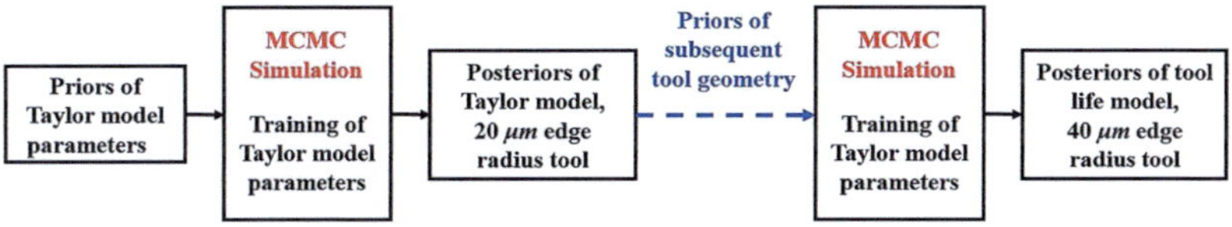

Figure 5.19: Sequential training and prediction of tool life using Bayesian updating for different tool edge radius

5.3.2.1 Establishing the Prior

Priors of the model parameters, C and n, are established to be the mean and standard deviation of posteriors of the 20 μm edge radius tool. Monte Carlo simulation is again used to find the functional form of the tool life prior; see Figure 5.20. The figure displays the prior median value and confidence intervals, and the experimental tool life data. Again, lognormal distribution was selected as the prior tool life function. The lower and upper bounds of the credible intervals were taken to be 15.87th and 84.13th percentiles, respectively. As can be seen in the figure, the prior median function overestimates the data points at the cutting speed of 300, 325, and 350 *m/min*.

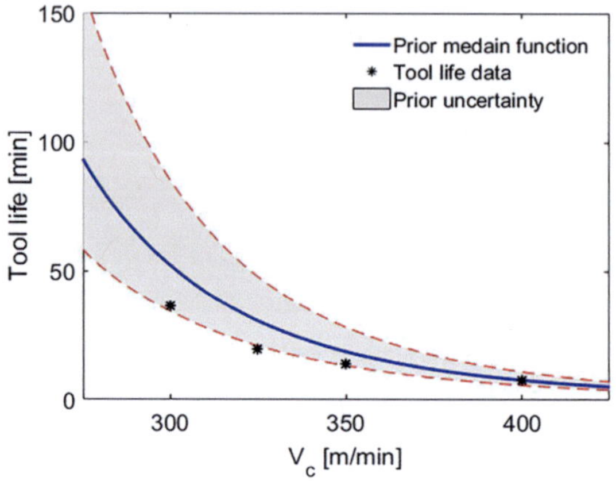

Figure 5.20: Prior function for the tool life using 40 μm edge radius tool

5.3.2.2 MCMC Simulation and Likelihood Function

Metropolis algorithm of MCMC simulation is used to update the priors of the parameters C and n using the tool life of 13.84 *min* at the cutting speed of 350 *m/min*. The number of iterations and burn-in period were selected to be $N = 10{,}000$ and 1000, respectively. To reduce the autocorrelation of the Markov chain, the thinning was implemented; see section 2.4.3. The acceptance ratio of the simulation was 0.33, denoting that 33% of the drawn samples are accepted by the MCMC simulation. Figure 5.21 displays the autocorrelation plots for 100 samples of the model parameters, C and n. As can be seen in the figures the autocorrelation of the chain converges to less than ± 0.02 after the 8th lag for both parameters.

Figure 5.22 depicts the joint posterior of the parameters C and n (right) after one update using the tool life 13.84 *min* at the cutting speed of 350 *m/min*. The mean values of the parameters C and n are 531 *m/min* and 0.158, and the parameters standard deviation values are 15.1 *m/min* and 0.01, respectively. Comparing the posterior and prior joint distributions, it is seen that the uncertainties are reduced. Additionally, the model parameters become correlated with the correlation coefficient of 0.85.

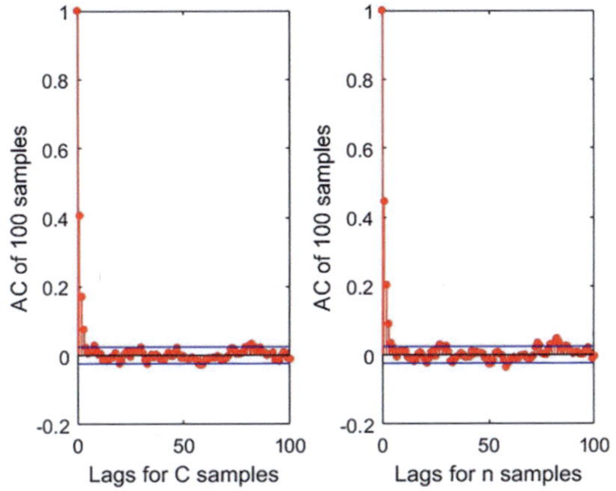

Figure 5.21: Autocorrelation for the C and n samples (converged chain)

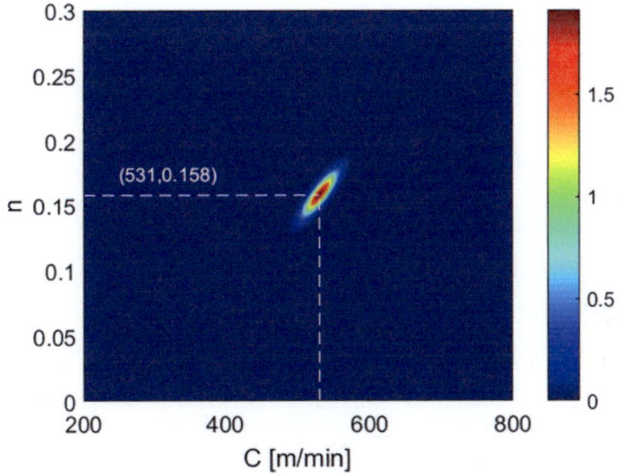

Figure 5.22: Posterior joint distribution of the parameters C and n for the 40 μm edge radius tool

5.3.2.3 Tool Life Prediction

Tool life posterior was found by inserting the posteriors C and n to the Eq. (5.1) using the Monte Carlo simulation. Figure 5.23 depicts the functional form of the tool life posterior with the mean and standard deviations of 2σ in addition to the training and test tool life data. The goodness of fit was computed

$R^2 = 0.977$. The tool life posterior was achieved using only one update due to the more informative prior. As can be seen, the posterior mean function represents the tool life data, accurately.

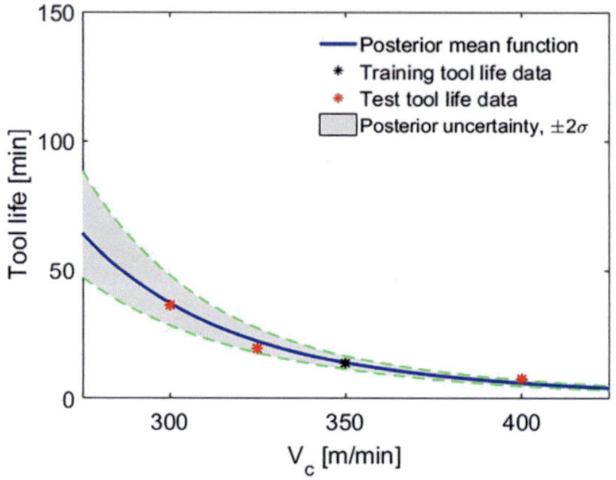

Figure 5.23: Posterior function with $\pm 2\sigma$ standard deviations uncertainty intervals using 40 μm edge radius tool

Bayesian inference was compared with the least squares curve fitting to evaluate the tool life prediction and the regression quality; see Figure 5.24. The regression fit quality using least square fitting was calculated to be $R^2 = 0.981$, which is slightly more accurate than the quality using Bayesian ($R^2 = 0.977$). However, the Bayesian approach requires only one data point for the parameters training. This demonstrates the effectiveness of Bayesian inference as compared to the least squares curve fitting, which requires at least two data points for the parameter identification, in this case.

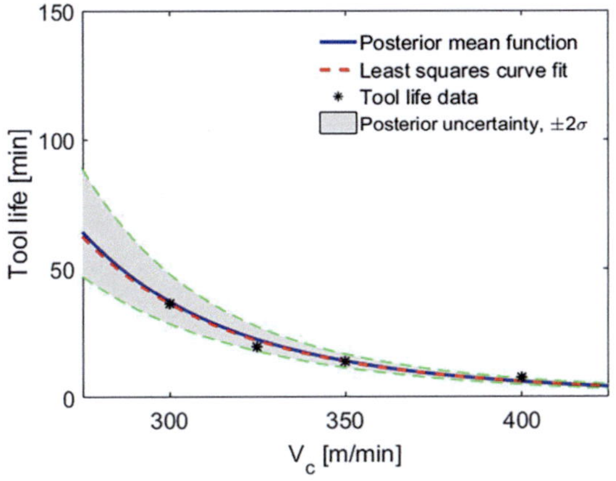

Figure 5.24: Comparison of posterior mean function and the least squares curve fitting methods using 40 *μm* edge radius tool

Figure 5.25 displays the lognormal prior (in blue color) and normal posterior (in red color) PDFs of the tool life and the measured life data at the cutting speed of 300 *m/min*. As illustrated in the figure, the uncertainty of the tool life is reduced, and the posterior can predict the tool life data, accurately.

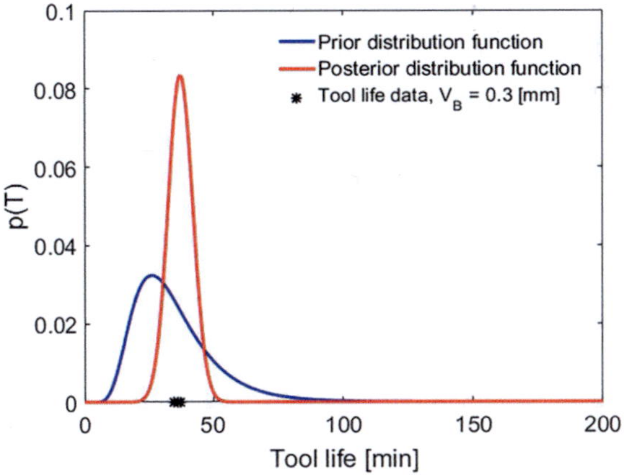

Figure 5.25: Prior and posterior PDFs using cutting speed 300 *m/min*

Table 5.5 lists the measured tool life and the predicted values using Bayesian inference for the tool edge radii 20 and 40 μm. The predicted values are provided with the mean and one standard deviation credible interval (σ). As can be seen all the tool life data points are predicted within two standard deviations (2σ). The minimum and maximum prediction errors for the tool life were calculated to be 0.7 and 21%, respectively. This implies that the model parameters identification and tool life prediction were performed with a good degree of accuracy using the MCMC method applied to the Taylor tool life model.

Table 5.5: Cutting conditions, tool life and Bayesian prediction for the tool edge radii 20 and 40 μm

No.	*Edge radius (μm)*	V_c *(m/min)*	$T_{\_measured}$ (min)	$T_{\_predicted}$ (min)	$T_{\_error}$ (%)
1	20	300	47.6	(47, 6.22)	1.2
2	20	325	28	(27.8, 2.68)	0.7
3	20	350	14.7	(17.1, 1.33)	16
4	20	400	7.7	(7.2, 0.69)	6
5	40	300	36.2	(37.2, 4.78)	2.7
6	40	325	19.5	(22.4, 2.38)	15
7	40	350	13.8	(13.9, 1.31)	0.7
8	40	400	7.6	(6, 0.85)	21

5.4 Reliability Analysis

Reliability function is the most frequently used function in engineering life data analysis. This function gives the probability of an item operating for a certain amount of time without failure. Eq. (5.3) represents the normal CDF of a tool failure, which refers to the probability that a tool is worn out prior or equal to the time *t*:

$$P(\tau < t) = \int_0^t f(\tau)\, d\tau \tag{5.3}$$

where τ is the time corresponding to the occurrence of tool failure. Note that this integral does not exist in a simple closed-form solution and it should be computed numerically. The reliability function is the complement of the CDF, which is expressed as,

$$R(t) = 1 - P(\tau < t) \tag{5.4}$$

The hazard function is the probability of failure at the time *(t, t+Δt)* given the unit has survived until time *t*. The function is defined as follows,

$$h(t) = \lim_{\Delta t \to 0} \frac{Pr(failure\ in\ interval(t, t + \Delta t)|no\ failure\ in\ (0, t))}{\Delta t} \tag{5.5}$$

$$h(t) = \frac{f(t)}{R(t)} \tag{5.6}$$

As illustrated in the Eq. (5.6), the hazard function is the probability distribution of tool life, *f(t)* over the reliability function, *R(t)*, at the time *t*. The hazard rate versus time plot is an important tool to understand how a cutting tool fails. If the rate decreases with time, the cutting tool exhibits pre-mature or early life failures. This type of failure is typically induced by mechanisms like design errors, poor quality control, the inadequacy of the tool to the applications, or material defects. If the hazard rate is constant with time, then the tool exhibits a random or memoryless failure rate behavior. If the failure rate is increasing with time, the cutting tool wears out (*i.e.,* mechanical wear or fatigue) [97,104]. Reliability analysis of the tool life considering the effect of the cutting speed and the tool geometries are described in the following sections.

5.4.1 Reliability Analysis for the Tool Edge Radius 20 and 40 *μm*

Figure 5.26 shows the reliability function of the tool life (using 20 *μm* edge radius tool) at the cutting speed of 300 *m/min* along with the measured tool life data points at the average flank wear of 0.3 (black points) and 0.35 *mm* (green points). As can be seen in the figure, the mean value of the posterior reliability function is predicted to be 46.9 *min*.

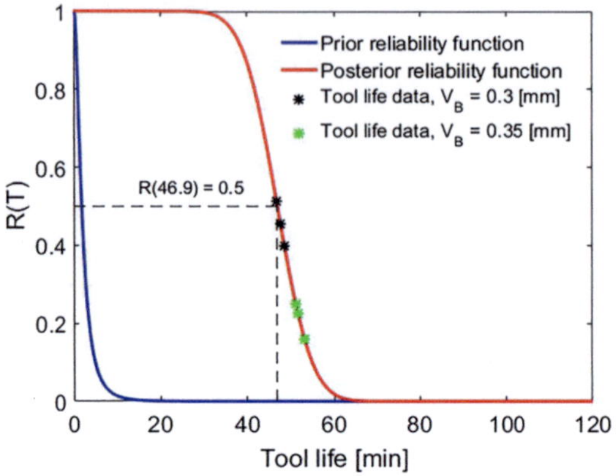

Figure 5.26: Prior and posterior reliability functions at cutting speed 300 *m/min* using 20 *μm* edge radius tool

Table 5.6 lists the tool life of the data points and the corresponding reliability values. According to the table, the reliability of the cutting tool drops significantly when the tool wear exceeds the average tool wear limit 0.3 *mm*.

Table 5.6: Measured tool life data and the corresponding reliability at cutting speed 300 *m/min* for the tool edge radius 20 *μm*

No.	*V_B (mm)*	*Tool life (min)*	*Reliability*
1	0.3	46.8	0.52
2	0.3	47.7	0.46
3	0.3	48.6	0.4
4	0.35	51.2	0.25
5	0.35	51.7	0.23
6	0.35	53.2	0.16

Since each of the reliability, probability density and hazard functions can completely represent the failure behavior of the tool at the corresponding cutting conditions, the hazard function can be used as a characteristic signature for qualitative performance evaluation; see Figure 5.27 [97].

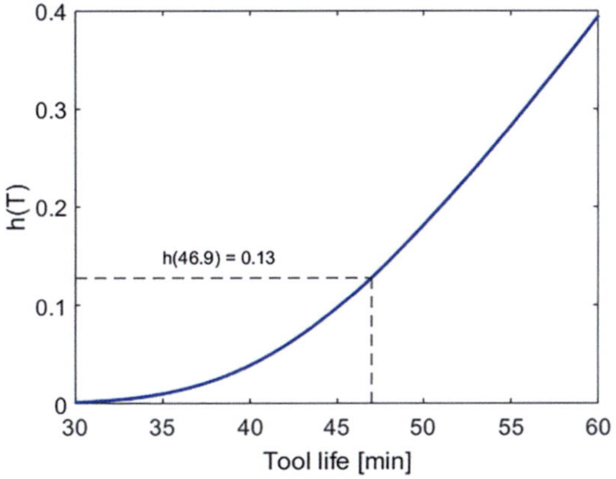

Figure 5.27: Posterior hazard function using cutting speed of 300 *m/min* using 20 *μm* edge radius tool

Reliability of cutting tool with edge radius of 40 *μm* can be calculated using the Eq. (5.4). Figure 5.28 illustrates the reliability function of the tool life at the cutting speed of 300 *m/min* along with the measured tool life data points at the average flank wear of 0.3 (black points) and 0.35 *mm* (green points). As can be seen in the figure, the mean value of the posterior reliability function is predicted to be 36.5 *min*.

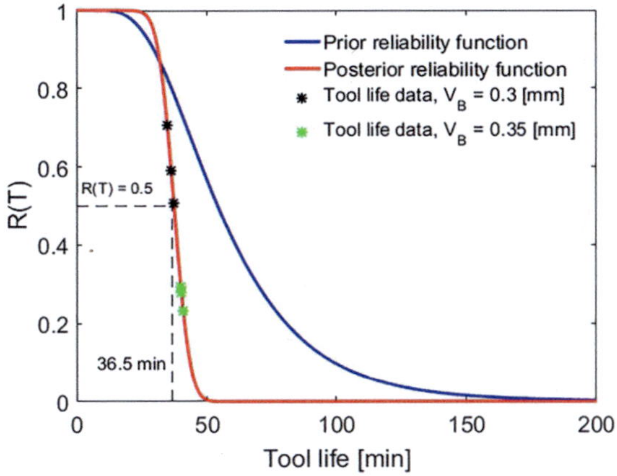

Figure 5.28: Prior and posterior reliability functions at cutting speed 300 *m/min* using 40 *μm* edge radius tool

Table 5.7 lists the tool life of the data points and the corresponding reliabilities. According to the table, the reliability of the cutting tool drops significantly when the tool wear exceeds the average tool wear limit 0.3 *mm*.

Table 5.7: Measured tool life data and the corresponding reliability at cutting speed 300 *m/min* for the tool edge radius 40 *μm*

No.	V_B *(mm)*	*Tool life (min)*	*Reliability*
1	0.3	34.7	0.71
2	0.3	36.2	0.6
3	0.3	37.2	0.51
4	0.35	39.9	0.3
5	0.35	40.1	0.28
6	0.35	40.8	0.23

5.4.2 Reliability Analysis Comparing Cutting Speeds

Figure 5.29 depicts the comparison between the posterior reliability functions at the cutting speeds of 350 and 400 *m/min* using the tool edge radius of 40 *μm*. As can be seen in the figure, the tool life reliability of 0.5 is calculated to be 13.9 and 6 *min* for the cutting speed of 350 and 400 *m/min*, respectively. The reliability of 0.5 denotes that 50% of the tools with the 40 *μm* edge radius can survive until the average flank wear, $V_{B,avg}$, of 0.3 *mm*.

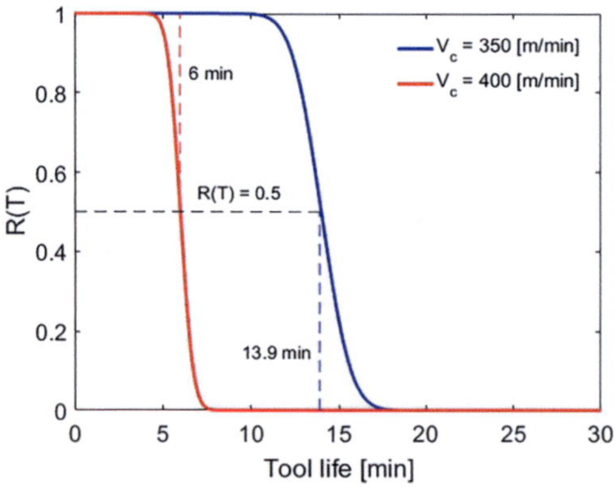

Figure 5.29: Comparison of posterior reliability functions using cutting speeds 350 and 400 *m/min*

While the probability density and reliability functions demonstrate the overall speed of failure, hazard function shows the dynamic (instantaneous) speed of failure. This provides a qualitative knowledge about the cutting tool failure supported by quantitative cutting tool life data.

Figure 5.30 demonstrates a comparison between the cutting tools hazard rate using the cutting speeds of 350 and 400 *m/min*, at the cutting time of 10 *min*. As illustrated, the instantaneous failure rates of the tools with an edge radius of 40 μm are 0.04 and 0.87 using the cutting speed 350 and 400 *m/min*, respectively. The hazard rate of the flank wear using the posterior normal distribution increase monotonically for both cutting speeds.

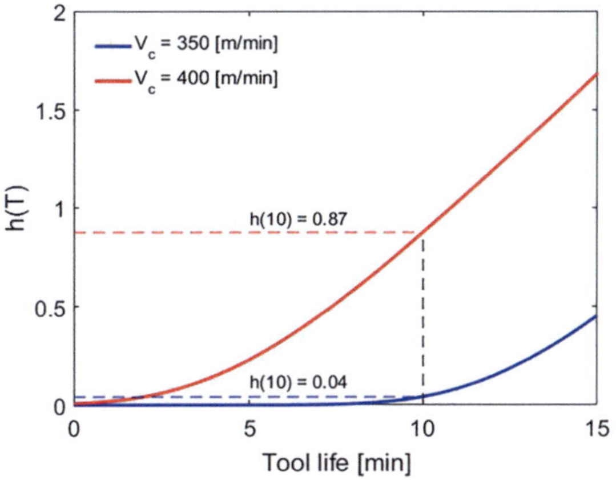

Figure 5.30: Comparison of posterior hazard functions using cutting speeds 350 and 400 *m/min*

5.4.3 Reliability Analysis Comparing Cutting Edge Geometries

Figure 5.31 depicts the comparison between the posterior reliability functions of the tool edge radii 20 and 40 μm at the cutting speed of 350 *m/min*. According to the figure, the reliability values of the posterior functions at the cutting time of the 15 *min* are 0.94 and 0.22 for the tool with edge radii 20 and 40 μm, respectively. This implies that the 20 μm edge radius tool can survive with the reliability of 94%, while the 40 μm edge radius tool survives with the reliability of 22%, until the cutting time of 15 *min*. This demonstrates that the former tool has more reliable performance and can be a preferred choice for the user in the defined machining application. Figure 5.32 demonstrates a comparison between the posterior hazard functions of the tool edge radii 20 and 40 μm at the cutting time of 16 *min*. As illustrated, the posterior hazard rates are 0.14 and 1.48 for the tool edge radii 20 and 40 μm, respectively. It is seen that the hazard function of 40 μm edge radius tool exceeds one.

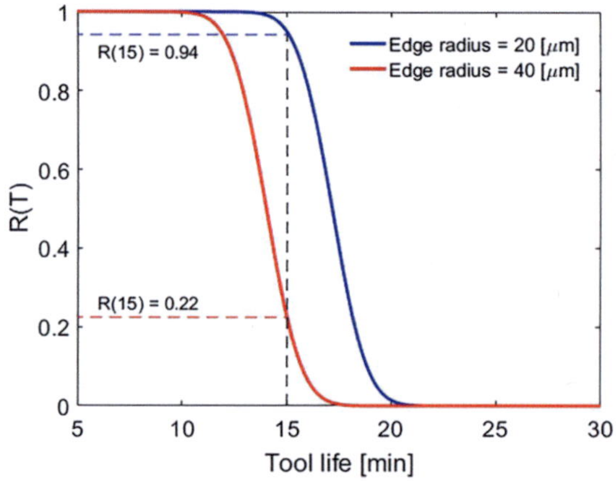

Figure 5.31: Comparison of posterior reliability functions for tool edge radii 20 and 40 μm at the cutting speed of 350 *m/min*

Essentially, the hazard rate is not a probability and can exceed one. In fact, it is the expected number of tool failure per unit of time, conditional on being at risk and not failed before the time, *t*. For this example, it is expected that the tool fails with the instantaneous rate of 1.48. Therefore, the tool with the smaller edge radius can minimize the risk of instantaneous failure compared to the bigger, 40 μm edge radius tool, in this case.

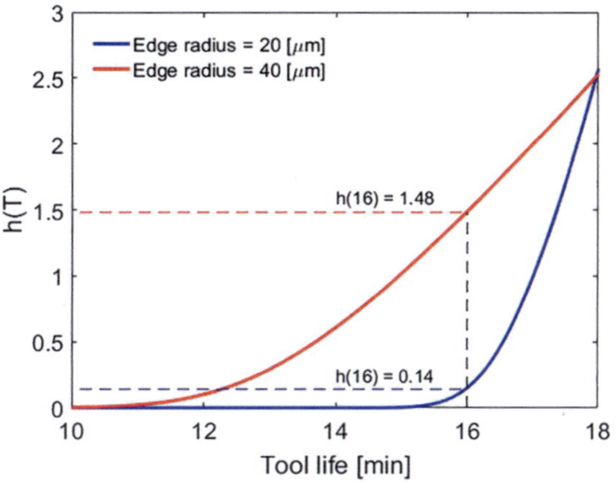

Figure 5.32: Comparison of posterior hazard functions for tool edge radii 20 and 40 *μm* at the cutting speed of 350 *m/min*

5.5 Conclusions

Probabilistic prediction and reliability analysis of tool life applying Bayesian MCMC to the Taylor tool life model were presented in this chapter. Sequential prediction of tool life using the tools with two different edge radii was performed successfully. In this regard, posteriors of the previous cutting speeds and tool geometries were used as priors of the current simulations, so that the tool life prediction was performed with the minimum input data. Lognormal distributions were used as prior PDFs, and the posterior normal PDFs were achieved after updating the model parameter using the Metropolis algorithm (*i.e.*, central limit theorem). The numerical quantification and minimization of the tool life uncertainty were conducted for the range of cutting speeds, 275-400 *m/min*, using MCMC simulation. It was also shown that the Bayesian method could predict the tool life for the second geometry using only one tool life data point thanks to the informative prior. This is impossible in the case of parameter determination by least squares curve fitting which requires at least two data points (in this case). The minimum and maximum prediction errors for the tool life were calculated to be 0.7 and 21%, respectively.

This implies that the model parameters identification and tool life prediction were performed with a good degree of accuracy using MCMC method applied to the Taylor tool life model.

The posteriors PDFs were used to calculate the reliability and hazard functions. Based on the quantitative and qualitative reliability analysis, it was demonstrated that:

1. The failure probability and instantaneous failure rate of the cutting tools (both geometries) are smaller using lower cutting speeds (*e.g.*, 350 *m/min*) compared to the cutting speed at 400 *m/min*.
2. The failure probability and instantaneous failure rate of the cutting tools with the edge radius of 20 μm are smaller than the failure functions using the tool with the of 40 μm edge radius tool.

This implies that machining with the lower cutting speeds and the smaller edge radius tools offer more reliable cutting tools usage and can minimize the risk of instantaneous failure. Nevertheless, reduction of cutting speed may affect the productivity and selection of smaller tool edge radius may cause vulnerability of the tool edge in case of heavy interrupted cuts. Therefore, the reliability analysis should be performed in the context of machining applications and take into account the applications limitations and user's preference.

6 Bayesian Updating for Tool Wear Growth Prediction in Milling Process

Probabilistic prediction of tool wear growth is presented in this chapter. Bayesian MCMC method is applied to identify the parameters of two models, exponential and Gompertz models. Three types of tools with the rake angles of 0, 3, and 9 *deg* as indexable milling tools were tested at the cutting speed value of 250 *m/min* until the tool flank wear threshold of 0.3 *mm* is reached. The model uncertainty is quantified and minimized numerically for all of the tool geometries. The probabilistic models are used for sequential prediction of tool wear growth using the information of previous tool geometry for subsequent tool geometries. This can reduce the cutting time required to identify tool wear evolution and the relevant end of life. The probabilistic models predict the tool wear curve and failure time using the degradation models and the wear data at the early and middle stages of the tool wear. The results of the probabilistic prediction of tool wear growth are validated with milling experiments data. Fit quality of the wear prediction (for the tool rake angle of 9 *deg*) using the Bayesian and least square methods are compared at the end.

6.1 Introduction

Tool wear curves illustrate the relationship between the amount of tool's flank wear, rake wear, and the cutting time (the cutting length). During the last decades, researchers have studied and modeled the tool wear using analytical, numerical and statistical models [105–108]. The models describe the wear rate as a function of some process variables (cutting velocity, feed rate) [43].

Usui *et al.* [109] have proposed an analytical wear model based on stress and the temperature in the interface between the cutting tool and workpiece. The model characterizes the wear rate with sensitive parameters of stress and temperature on the tool face. The measurements of the stress and the temperature must be performed and insert to the wear model. However, the experiments with the dedicated setup could not be performed practically in the machine shops. Therefore, they have improved the wear rate model, in which model constants could be identified using cutting tests without specialized equipment [110]. The model (Eq. (6.1)) is characterized by normal stress, σ_f, temperature on the flank wear land, θ_f, cutting speed, V_c, and tool rake angle, α, relief angle, γ, and two constants C and λ.

$$\frac{\mathrm{dV}_B}{\mathrm{dt}} = C\,\sigma_f \exp\left(-\frac{\lambda}{\theta_f}\right)\cdot\left(\frac{1}{\tan\gamma} - \tan\alpha\right)V_c \tag{6.1}$$

Takeyama and Murata [111] derived a fundamental wear rate equation by considering abrasive wear which is proportional to cutting distance as a function of the activation energy of the diffusion process. Since the Usui's [109] and Murata's [111] wear models are derived as a function of the process parameters (*e.g.*, tool temperature, contact pressure and sliding velocity of the chip), they are widely used to be implemented in the FEM codes [112]. 2D FEM codes implementing tool wear models have been reported in the literature [113,114]. However, their most significant limitation is that they are able to predict the tool wear only for orthogonal cutting conditions. 3D FEM simulations have been implemented, using the Deform 3D environment, to predict the tool wear in longitudinal turning operation [115,116]. The developed models can consider tool geometry modification due to the wear rate and how its distribution changes as the tool geometry changes.

Among the statistical methods, artificial neural network and regression analysis have also been used for the tool wear modeling. Karpat *et al.* [117] have used predictive neural network modeling to predict tool wear in finish hard turning process using Cubic Boron Nitride (CBN) tools. Attanasio *et al.* [118] compared response surface methodology (RSM) and artificial neural networks (ANNs) fitting techniques for prediction of tool wear for turning of AISI 1045 steel. The comparison showed that ANNs model provides better approximation than RSM in the prediction of the amount of the tool wear parameters. Klocke *et al.* [119] proposed a regression analysis to determine the material constants C and λ in Usui's wear model (Eq. (6.1)). The abrasive model can predict both the flank and the crater wear, in which the experimental data points can be interpolated by only one line in a semi-logarithmic chart.

In this chapter, probabilistic prediction of tool flank wear is presented. The probabilistic model can take into account the variability of machining process parameters and quantify the relative inherent uncertainty. The Bayesian MCMC is used to calculate posterior distributions of the degradation models parameters-exponential and Gompertz models. The Bayesian method is used to incorporate the prior knowledge (about model parameters from previous experiments or user belief) into the tool wear analysis which leads to minimizing the number of experiments required for the model parameters identification.

Figure 6.1 shows the mean and one standard deviation error bars of the tool flank growth as a function of cutting time at the cutting speed of 250 *m/min* and feed of 0.05 *mm/tooth* using a tool with the rake angle of 9 *deg*. As can be seen, the tool wear growth is divided into three zones described as follows:

1. **Break-in period:** It refers to rapid initial wear which occurs in the early cutting time.
2. **Steady-state wear region:** This region denotes a uniform tool wear rate.
3. **Failure region:** It indicates an accelerating wear rate until final failure occurs.

The goal of this work is to train the model parameters of the degradation models (exponential and Gompertz) and predict the tool wear growth, probabilistically. In this regard, the break-in period is discarded from the analysis, and the model parameters are trained to predict the tool wear curve in the steady-state and failure regions. The training of the model parameters is performed as following steps:

1. For the first tool geometry (0 *deg* tool rake angle) the model's parameters updating is done using wear data of the steady-state and failure region regions.
2. For the second and third tool geometries (3 and 9 *deg* tool rake angles), posteriors of the previous tool geometries are used as priors of the subsequent geometries. The sequential probabilistic prediction and updating are performed using the wear data of the steady-state region.

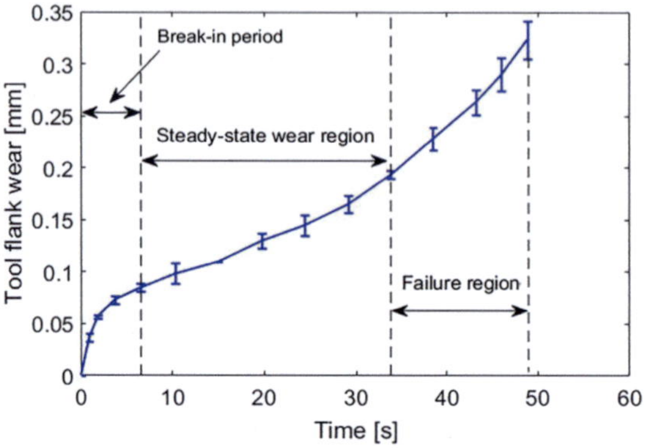

Figure 6.1: Tool flank wear growth as a function of cutting time divided into three regions

The sequential probabilistic technique allows to predict the end of tool wear criterion (in this case, tool wear amount of 0.3 *mm*) taking into account the early or middle stage of tool wear data. In this way, tool wear tests are not required to be conducted until the end of the wear criterion, so that the tool end of life can be determined beforehand. In addition, using the probabilistic modeling technique, one can predict the tool wear growth with mean and standard deviation uncertainty intervals for each wear data point.

The chapter is organized as follows. Section 6.2 reports the experimental setup and results of the milling tests. Section 6.3 presents the application of MCMC to the exponential model for sequential prediction of tool wear growth. Section 6.4 demonstrates the application of MCMC to the Gompertz model for sequential prediction of tool wear growth. Discussions and conclusions are presented in chapters 6.5 and 6.6.

6.2 Experimental Setup, Results, and Discussion

Milling experiments were performed on a 5-axis milling machine Hermle C40; see Figure 5.2. Slot milling tool wear tests (2.5 axes) were completed using three face mills with a diameter of 40 *mm*, holding three cutting inserts. The face mills were produced in three different rake angles (0, 3, and 9 *deg*). Cutting inserts were selected to be uncoated (SPGW09T308) Tungsten carbide with the edge radius of 40 *µm*. The inserts were designed and produced by Zermet Zerspanung GmbH, with the ISO grade of P25. The cubic workpiece material was AISI 1045 steel with the dimensions of 120×120×100 *mm*. The depth of cut was selected to be 1.5 *mm*. Feed value of 0.05 *mm/tooth* and a cutting speed of 250 *m/min* were selected for the tests.

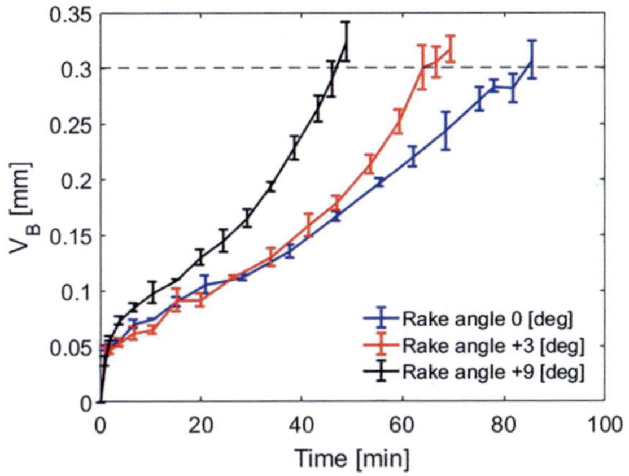

Figure 6.2: Comparison of tool flank wear for tools with three different rake angles at a cutting speed of 250 m/min and feed of 0.05 *mm/rev*

Figure 6.2 shows the tool flank wear growth comparison of three types of tool rake angles (0, 3, and 9 *deg*) at the cutting speed of 250 m/min and the feed of 0.05 *mm/rev*. As can be seen, the tool life decreases with the increase in tool rake angles. In other words, the tool with the largest rake angle (9 *deg*) reaches the tool wear criterion, $V_B = 0.3$ *mm*, faster than others. Table 6.1 lists the tool life values of the tool rake angles at $V_B = 0.3$ *mm*.

Table 6.1: Measured tool life for various tool rake angles

No.	V_B *(mm)*	*Tool rake angle (deg)*	$T_{\_measured}$ *(min)*
1	0.3	0	83.2
2	0.3	3	63.8
3	0.3	9	46.8

6.3 Probabilistic Prediction of Tool Wear Growth using Exponential Function

Probabilistic prediction of tool wear growth is performed applying MCMC method to model parameters of the exponential function. The function is often used to model destructive degradation of the components, which is given as:

$$y = b \,.\, e^{a.t} \tag{6.2}$$

where y is the mean value of the tool wear for the corresponding measured time t. a and b are the model's parameters, which are required to be determined. The normal distribution is used to represent the mean and uncertainty of the tool wear data points.

In the exponential model, there is uncertainty in the wear value, y, due to the uncertainties in the model parameters, a, and b. The Blockwise MCMC is used to estimate the exponential model parameters. Metropolis algorithm is again used to draw samples from the joint proposal distribution, $q(a,b)$, to approximate the posterior target distribution, $p(a,b)$; see Algorithm 6.1:. According to the algorithm, the likelihood function calculates the probability of the measured tool life, V_B, given the joint model parameters, (a,b). The posterior distribution is calculated by multiplying the prior joint distribution into the likelihood function.

To develop the sequential probabilistic models about the model parameters of the exponential function, the parameters priors are established for the 0 *deg* rake angle tool, first. Second, the priors are updated using the wear tests data to obtain the posterior distributions of the parameters and tool wear curve. Third, the results of the parameters posteriors are used as priors of the tool rake angle 3 *deg*. The sequential modeling technique is continued to predict the tool wear curve for the 9 *deg* rake angle tool; see Figure 6.3.

Algorithm 6.1: Metropolis algorithm for updating of the exponential function parameters

1. Establish a normal prior distribution, $p(a,b)$,
2. Establish a proposal density function for (a,b),
3. Initialize a starting sample $(a,b)^0$,
4. For i = 0 to i = N-1:
- Select a candidate $(a,b)^{new}$ from a proposal distribution, $q((a,b)^{new} | (a,b)^i)$,
- Compute the posterior distribution, $p((a,b)^i | V_B) = p((a,b)^i)\, p(V_B \mid (a,b)^i)$,
- Calculate the acceptance ratio, $r = \frac{p((a,b)^{new})}{P((a,b)^i)}$,
- Generate a random number, $u \sim uniform\ (0,1)$,

    If $u \leq r$:

        Accept the proposal: $(a,b)^{i+1} = (a,b)^{new}$,

    Else:

        Reject the proposal: $(a,b)^{i} = (a,b)^{new}$,

    End If

5. End For

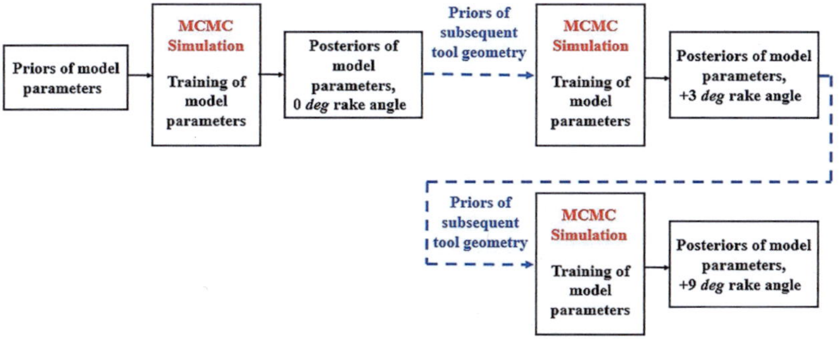

Figure 6.3: Sequential probabilistic prediction of the tool wear growth using exponential and Gompertz models

6.3.1 Wear Growth Prediction using Tool Rake Angle 0 *deg*

The parameters identification starts with establishing prior values for the parameters, a and b, of the exponential function. Since there is no information available for the model parameters, the priors were selected as uniform distributions. In the Bayesian analysis, a uniform distribution is referred to as a non-informative distribution, where the probability of the random variables is distributed equally likely within the specified range. The ranges of the uniform priors (for a and b parameters) were selected to be between 0 and 2. As can be seen in Figure 6.4, the probability of the distribution is 0.5 through the entire range of the parameter, a. Priors of the parameters, a and b, are updated with the results of the measured tool wear (likelihood function) to obtain the posterior distribution.

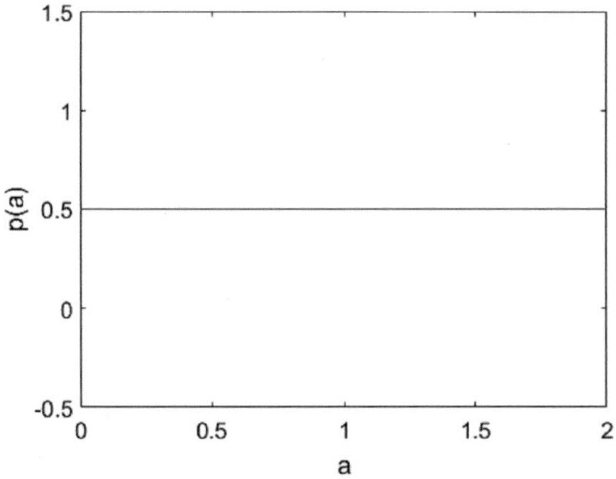

Figure 6.4: Uniform prior distribution for the parameter a

The likelihood function using the wear data of the 0 *deg* rake angle tool is given as follows:

$$p(V_B|a,b) = e^{-\frac{((b\,.e^{a.t})-V_B)^2}{2\sigma_{V_B}{}^2}} \tag{6.3}$$

where $p(V_B| a,b)$ is the likelihood function of the measured average tool wear, V_B, given specified prior values of the joint parameters distribution, (a,b), at a specific cutting time. The likelihood function is expressed as a non-normalized normal distribution, where σ_{VB} is the standard deviation of the measured tool wear. The standard deviation for this case was selected to be 5-7% of the measured tool wear. The likelihood function describes how likely the measurement result at a particular time is, given prior parameters values. MCMC simulation were implemented to calculate the posterior distributions of the parameters. The Metropolis algorithm was carried out for N = 10,000 iterations and 1500 samples were taken as the burn-in period. Figure 6.5 displays the joint posterior distribution, *(a,b)*, after training of the parameters using the tool wear data until the wear criterion, 0.3 *mm*. Joint Gaussian distribution is used to illustrate the posteriors. The mean values of the parameters a and b are 0.066 *m/min* and 0.018, and the standard deviation values are 0.007 *m/min* and 0.002, respectively. Moreover, the uncertainty of the posterior function is reduced. As can be seen, the model parameters become correlated with the correlation coefficient of -0.85 for *(a,b)* joint distribution.

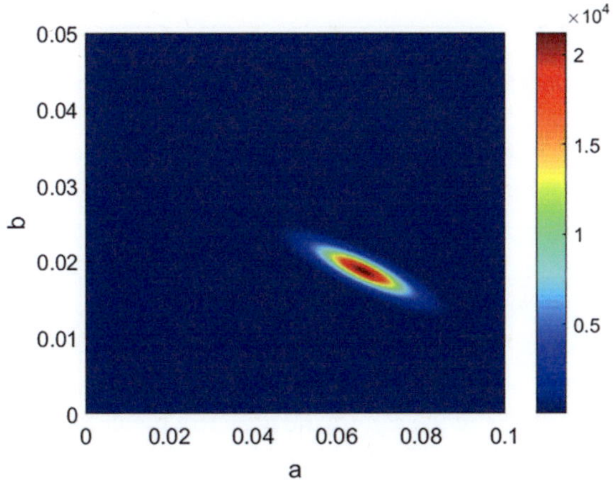

Figure 6.5: Joint posterior distribution of the parameters a and b for the tool rake angle 0 *deg*

Figure 6.6 displays marginal posterior distribution of the parameter, *a*, and the corresponding trace plot for the $N = 10{,}000$ iterations after discarding the burn-in period. The sampled distribution is fitted by a normal distribution, accurately. As demonstrated, the Markov chain converges to a stationary condition within the drawn samples range.

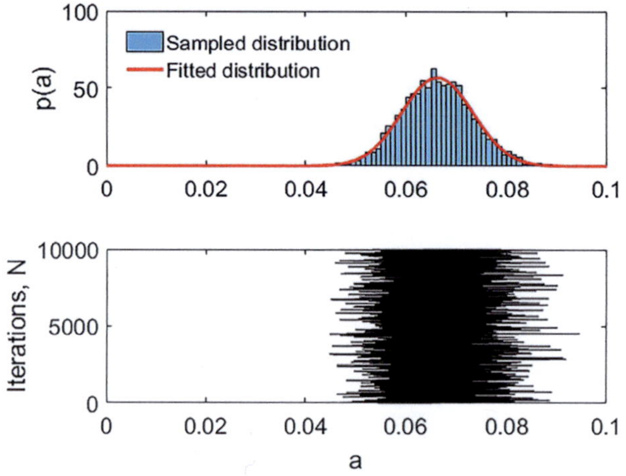

Figure 6.6: Marginal posterior distribution and trace plot of the parameter *a*

The posterior exponential function of the tool wear is obtained using Monte Carlo simulation by inserting the marginal posterior, *a* and *b*, to the Eq. (6.2). Figure 6.7 depicts the posterior mean function with 2σ standard deviations uncertainty intervals to approximate the tool wear curve of 0 *deg* rake angle tool. The uncertainty of the posterior function was computed, $\sigma_{VB} = 0.009$ *mm*, using Monte Carlo simulation at the time 37.5 *min*, which is the middle-measured data point of the tool wear curve. The regression fit quality was calculated to be $R^2 = 0.987$. According to the figure, the posterior mean function predicts the end of wear criterion (at the wear value of 0.3 *mm*), 80.62 *min*. As can be seen, the measured flank wear with the error bars (including the end of tool wear data point) appears within the credible intervals of the posterior function. It is important to note that the posterior function predicts the tool wear at the steady-state and failure region (are marked with the pink dash lines) and not the break-in period.

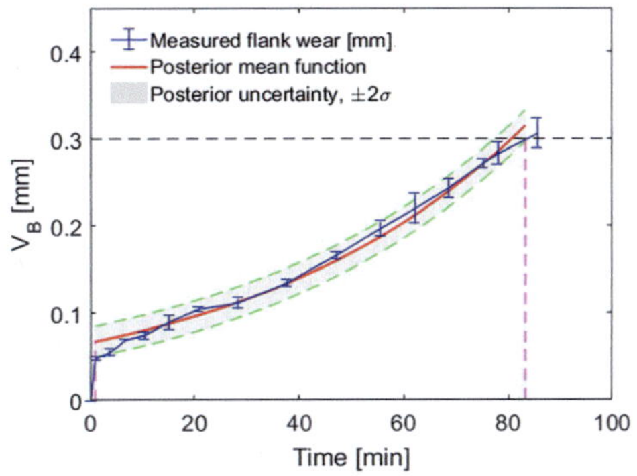

Figure 6.7: Posterior exponential function with $\pm 2\sigma$ standard deviations uncertainty intervals versus the measured tool wear data using the tool rake angle 0 *deg*

6.3.2 Sequential Wear Growth Prediction

Sequential prediction of the tool wear growth is performed using posterior mean values of the parameters, *a* and *b*, for the 0 *deg* rake angle tool as priors of the subsequent tool geometry 3 *deg* tool rake angle; see Figure 6.3. The prior joint distribution of the parameters for the 3 *deg* rake angle tool was taken to be independent. The standard deviations of the priors were selected to be 0.01, which is bigger than the standard deviations, obtained using previous tool geometry. This is due to the fact that allocating bigger uncertainties to the priors enables the simulations to rely more on measurements (*i.e.*, less confidence in the prior knowledge).

Figure 6.8 shows the prior exponential function with 2σ standard deviations uncertainty intervals using the tool rake angle 3 *deg*. Once again, the Monte Carlo simulation is used to compute the prior function. Although the prior function is informative (in comparison to the uniform distribution), its mean function under-estimates the measured tool wear. The advantage of an informative prior is that it influences the posterior distribution and is not entirely dominated by likelihood function. Therefore, fewer data points are required for updating the function to achieve posterior probabilities. The proper use of prior distributions illustrates the power of the Bayesian method, in which the information is gathered from the previous study, past experience, and expert opinion so that the information can be naturally combined into current analysis.

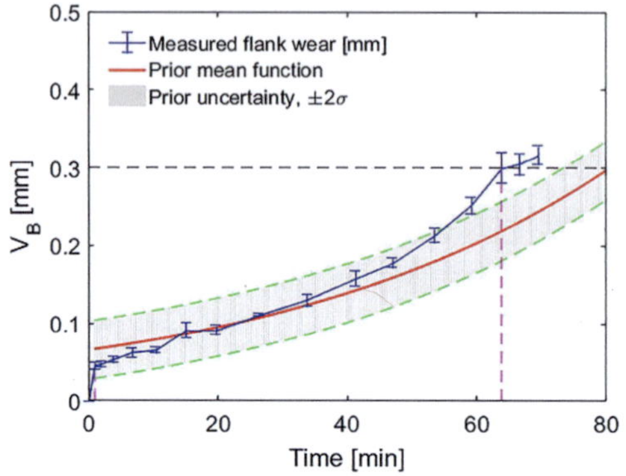

Figure 6.8: Prior exponential function with $\pm 2\sigma$ standard deviations uncertainty intervals versus the measured tool wear data using the tool rake angle 3 *deg*

MCMC simulation is practiced to calculate the posterior distribution of the parameters, *a* and *b*. Figure 6.9 displays the joint prior (left) and posterior (right) of the parameters after updating process with the tool wear data until the cutting time 33.78 *min* (measured data point at the middle of the tool wear curve). As can be seen, the uncertainty of the posterior distribution is reduced compared to the prior distribution. The mean values of the parameters *a* and *b* were obtained to be 0.054 *m/min* and 0.026 and the standard deviation values were calculated to be 0.003 *m/min* and 0.002, respectively. As can be seen, the model parameters become correlated with the correlation coefficient of -0.91 for *(a,b)* joint distribution.

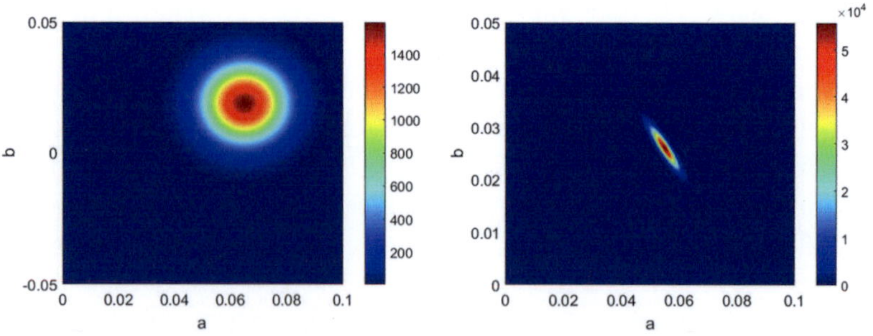

Figure 6.9: Prior (left) and posterior (right) joint Gaussian distribution of the parameters using the tool for the 3 *deg* tool rake angle

The posterior exponential function of the tool wear for the 3 *deg* rake angle tool is obtained using Monte Carlo simulation by inserting the marginal posteriors, *a* and *b*, to the Eq. (6.2). Figure 6.10 displays the posterior mean function with 2σ standard deviations credible intervals to predict the measured tool wear curve. The uncertainty of the posterior function was quantified, was computed, $\sigma_{VB} = 0.012$ *mm*, using Monte Carlo simulation at the time 26.27 *min*, which is the middle-measured data point of the tool wear curve.

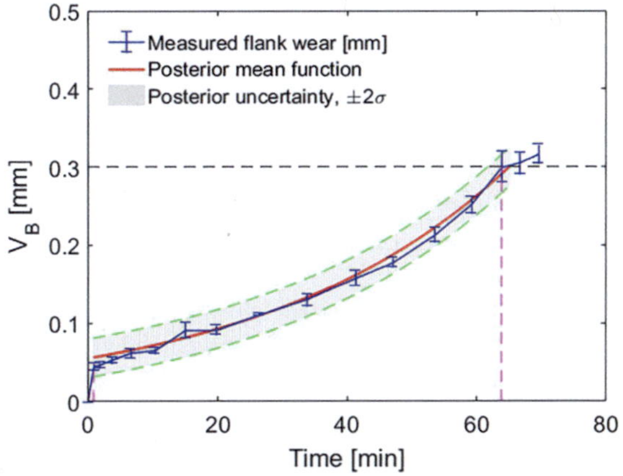

Figure 6.10: Posterior exponential function with $\pm 2\sigma$ standard deviations uncertainty intervals versus the measured tool wear data using the 3 *deg* rake angle tool

The regression fit quality was calculated to be $R^2 = 0.991$. According to the figure, the posterior mean function predicts the end of wear criterion, 65.06 *min*. As illustrated in the figure, the measured wear and the error bars are all happened to be within the credible intervals of the posterior function until the tool wear, $V_B = 0.3$ *mm*. The posterior function estimates the tool wear at the steady-state and failure region.

Figure 6.11 shows the prior exponential function with 2σ standard deviations uncertainty intervals using the tool rake angle 9 *deg*. As demonstrated, the prior under-predicts the measured tool wear.

Metropolis MCMC simulation is again executed to calculate the posterior distribution of the parameters, *a* and *b*. Figure 6.12 shows the joint prior (left) and posterior (right) of the parameters after the updating process with the tool wear data until the cutting time 24.4 *min* (*i.e.*, the middle wear data point of the tool wear curve). As demonstrated in the figure, the uncertainty of posterior distribution is minimized after the parameters training process. The mean values of the parameters *a* and *b* are 0.068 m/min and 0.031, and the standard deviation values are 0.004 m/min and 0.003, respectively. Although the prior joint PDF was selected to be independent, the joint posterior PDF becomes correlated with the correlation coefficient of -0.93 for *(a,b)* joint distribution.

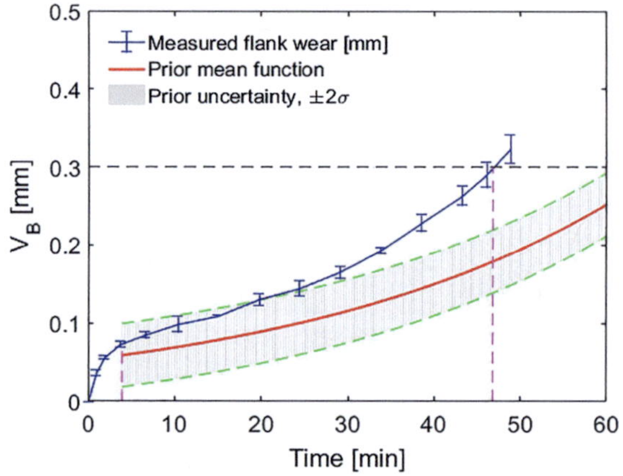

Figure 6.11: Prior exponential function with $\pm 2\sigma$ standard deviations uncertainty intervals versus the measured tool wear data using the tool rake angle 9 *deg*

Metropolis MCMC simulation is again executed to calculate the posterior distribution of the parameters, *a* and *b*. Figure 6.12 shows the joint prior (left) and posterior (right) of the parameters after the updating process with the tool wear data until the cutting time 24.4 *min* (*i.e.*, the middle wear data point of the tool wear curve). As demonstrated in the figure, the uncertainty of posterior distribution is minimized after the parameters training process. The mean values of the parameters *a* and *b* are 0.068 m/min and 0.031, and the standard deviation values are 0.004 m/min and 0.003, respectively. Although the prior joint PDF was selected to be independent, the joint posterior PDF becomes correlated with the correlation coefficient of -0.93 for *(a,b)* joint distribution.

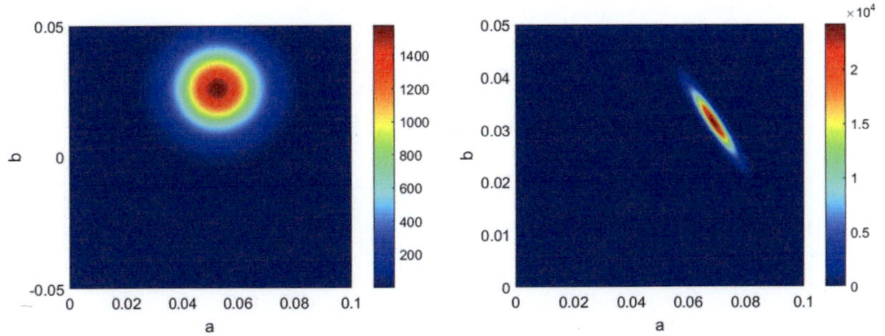

Figure 6.12: Prior (left) and posterior (right) joint Gaussian distribution of the parameters using the tool for the 9 *deg* tool rake angle

The posterior exponential function of the tool wear for the 9 *deg* rake angle tool is obtained using Monte Carlo simulation by inserting the marginal posteriors, a and b, to the Eq.(6.2).

Figure 6.13 displays the posterior mean function with 2σ standard deviations credible intervals to predict the measured tool wear curve. The uncertainty of the posterior function was evaluated, σ_{VB} = 0.008 *mm*, using Monte Carlo simulation at the time 24.4 *min*. The regression fit quality was calculated to be R^2 = 0.997. The posterior mean function predicts the end of wear criterion, 46.9 *min*. As illustrated in the figure, the measured wear and the error bars appear within the credible intervals of the posterior function until the end of wear value, $V_B = 0.3$ *mm*.

The Bayesian method is compared with the deterministic approach to compare the assess the fit quality and accuracy of the tool life prediction at $V_B = 0.3$ *mm*. Figure 6.14 shows the prediction of the tool wear curve using the least squares fitting method. The model parameters were trained with the same wear data points, which were used for training of the Bayesian model. The regression fit quality using the least squares method was calculated to be $R^2 = 0.971$. According to the figure, the function cannot accurately predict the tool wear curve in the failure region, where the predicted end of wear criterion is 49.8 *min*.

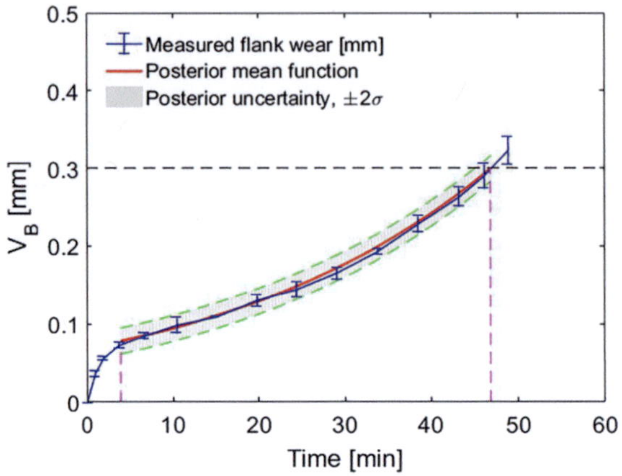

Figure 6.13: Posterior exponential function with $\pm 2\sigma$ standard deviations uncertainty intervals versus the measured tool wear data using the tool rake angle 9 *deg*

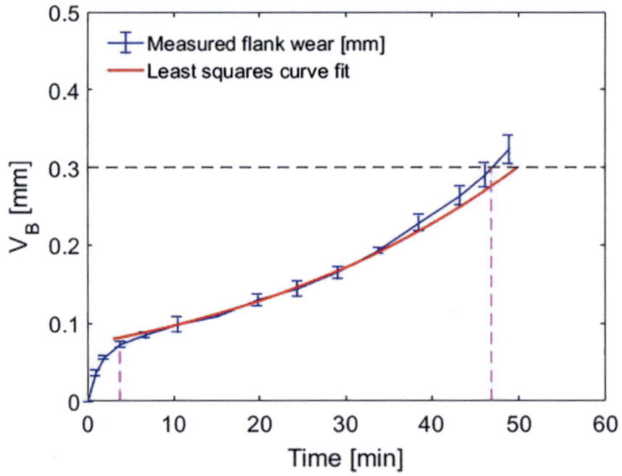

Figure 6.14: Exponential function using least squares fit versus the measured tool wear data for the 9 *deg* rake angle tool

Table 6.2 compares the tool life prediction error and regression quality using both methods for the 9 *deg* rake angle tool. As can be seen, the Bayesian method is able to approximate the tool wear curve with better regression quality. Moreover, the error percentage values of the tool life prediction were computed to be 0.2, and 6.5% for the Bayesian and least squares fit methods, respectively.

This shows that the Bayesian method can estimate the tool life more accurate than the least squares fit method.

Table 6.2: Comparison of Bayesian and least squares methods using exponential function

Method	*Tool rake angle (deg)*	$T_{\_measured}$ *(min)*	$T_{\_predicted}$ *(min)*	$T_{\_error}$ *(%)*	$R_{\_Squared}$
Bayesian	9	46.8	46.9	0.2	0.997
Lease squares fitting	9	46.8	49.8	6.5	0.971

6.4 Prediction of Tool Wear Growth using Gompertz Model

MCMC method is applied to the model parameters of the Gompertz function to predict the tool wear curve. The function is often used to model the non-destructive and destructive degradation tests of the components of the machine. A deterministic form of the function is written as:

$$y = a \,.\, b^{c^t} \tag{6.4}$$

where y is the mean value of the tool wear for the corresponding measured time t. a, b and c are the model's parameters, which can be identified using Bayesian and least squares fitting methods. The normal distribution is again used to represent the mean and uncertainty intervals of the tool wear data points. The sequential probabilistic modeling of the tool wear curve is practiced applying the MCMC method to the Gompertz model; see Figure 6.3. In this context, the model parameters of the Gompertz function are identified for the 0 *deg* rake angle tool, first. Second, the results of the parameters posteriors are then used as priors of the tool rake angle 3 *deg*. The sequential probabilistic modeling technique is continued to predict the tool wear curve for the 9 *deg* rake angle tool.

Since the sequential probabilistic modeling using the Gompertz model is identical to the exponential function, the procedure for model parameters identification is not reported for the sake of brevity. Hence, only the results of the tool wear curve prediction are shown for the tool rake angles 0, 3, and 9 *deg* in this section. Figure 6.15 displays the posterior mean of the Gompertz function with 2σ standard deviations uncertainty intervals to estimate the wear curve of 0 *deg* rake angle tool. The uncertainty of the function was quantified, $\sigma_{VB} = 0.007$ *mm*, using Monte Carlo simulation at the time 37.5 *min*, at the middle-measured data point of the tool wear curve. The regression fit quality was calculated to be $R^2 = 0.984$. According to the figure, the posterior mean function predicts the end of wear criterion (at the wear value of 0.3 *mm*), 81.95 *min*.

According to the figure, the measured wear with the error bars (including the tool life at $V_B = 0.3$ *mm*) appears within the credible intervals of the posterior function. Once again, the posterior function estimates the tool wear at the steady-state and failure region and not the break-in period.

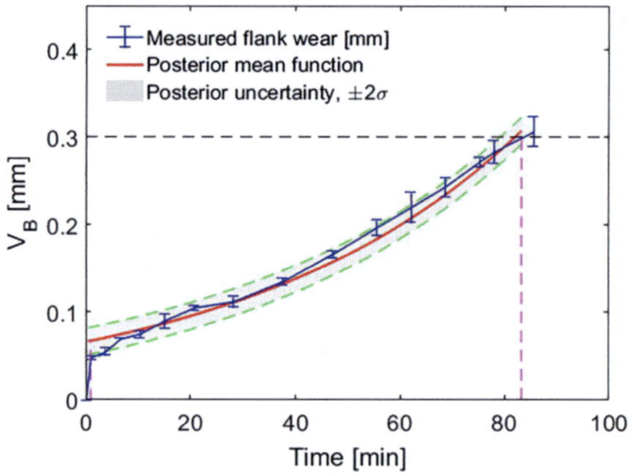

Figure 6.15: Posterior Gompertz function with $\pm 2\sigma$ standard deviations uncertainty intervals versus the measured tool wear data using the 0 *deg* rake angle tool

The sequential tool wear prediction is repeated using the posterior of the previous geometry as priors of the new geometry. Figure 6.16 displays the posterior Gompertz mean function with 2σ standard deviations credible intervals to estimate the measured tool wear curve using the 3 *deg* rake angle tool. The uncertainty of the posterior function was computed, $\sigma_{VB} = 0.007$ *mm*, using Monte Carlo simulation at the time 26.27 *min* (which is the middle-measured data point of the tool wear curve).

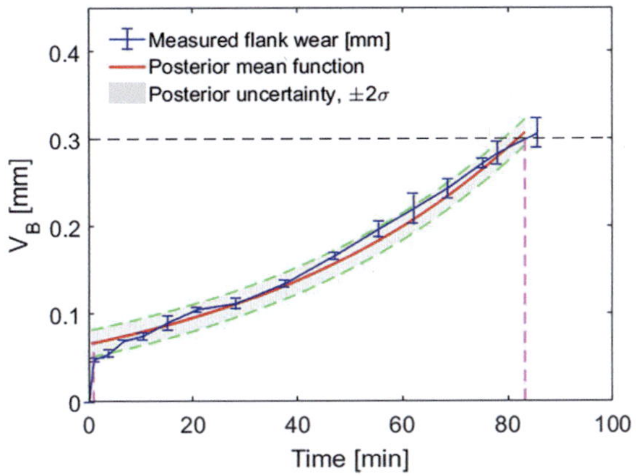

Figure 6.16: Posterior Gompertz function with $\pm 2\sigma$ standard deviations uncertainty intervals versus the measured tool wear data using the tool rake angle 3 *deg*

The regression fit quality was achieved to be $R^2 = 0.988$. According to the figure, the posterior mean function predicts the end of wear criterion, 64.4 *min*. Additionally, the measured wear and the corresponding error bars are all happened to be within the 2σ uncertainty intervals of the posterior function until the tool wear amount of $V_B = 0.3$ *mm*. The posterior function estimates the tool wear at the steady-state and failure region.

Figure 6.17 shows the posterior Gompertz mean function with 2σ standard deviations uncertainty intervals to estimate the measured tool wear curve using the 9 *deg* rake angle tool. The uncertainty was quantified, $\sigma_{VB} = 0.005$ *mm*, using Monte Carlo simulation at the time, 24.4 *min*. The regression fit quality was computed to be $R^2 = 0.997$. The posterior mean function predicts the end of wear criterion, 47.2 *min*. As demonstrated in the figure, the measured wear and the error bars appear within the credible intervals of the posterior function in both steady-state and failure regions.

The Bayesian method is compared with a deterministic approach to evaluate the regression fit quality and accuracy of the tool life prediction at $V_B = 0.3$ *mm*. Figure 6.18 shows the prediction of the tool wear curve using the least squares fitting method. The model parameters were identified with the same wear data points, which were used for training of the Bayesian model.

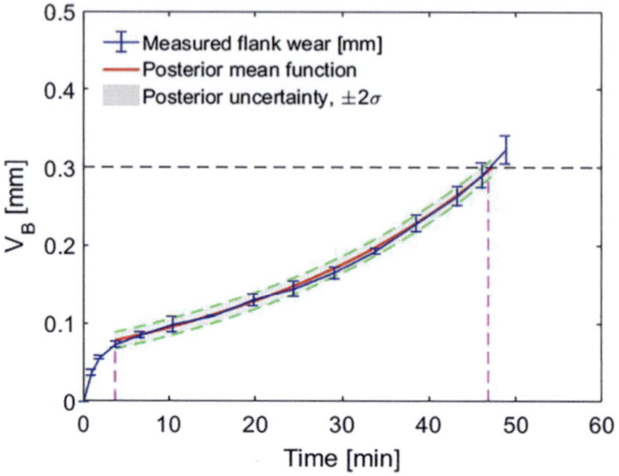

Figure 6.17: Posterior Gompertz function with $\pm 2\sigma$ standard deviations uncertainty intervals versus the measured tool wear data using the tool rake angle 9 *deg*

The regression fit quality was calculated to be $R^2 = 0.971$. The least squares function estimates the tool life, 49.8 *min*. According to the figure, the function is not able to predict the tool wear curve in the failure region, accurately.

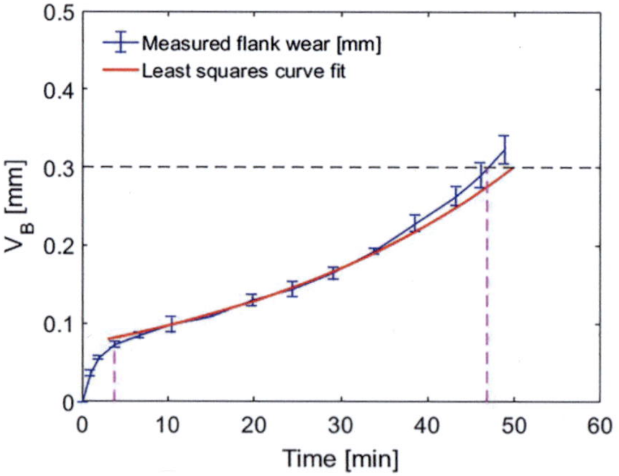

Figure 6.18: Gompertz function using least squares fit versus the measured tool wear data for the tool rake angle 9 *deg*

Table 6.3 compares the tool life prediction error using the posterior mean function and regression quality using both methods for the 9 *deg* rake angle tool. As can be seen, the Bayesian method is able to approximate the tool wear curve with better regression quality. Moreover, the error percentage values of the predicted tool life were computed to be 0.8 and 6.5% for the Bayesian and least squares fit methods, respectively. Again, the Bayesian method can estimate the tool life more accurate than the least squares fit method using the Gompertz model.

Table 6.3: Comparison of Bayesian and least squares methods using Gompertz function

Method	*Tool rake angle (deg)*	$T_{\_measured}$ *(min)*	$T_{\_predicted}$ *(min)*	$T_{\_error}$ *(%)*	$R_{\_Squared}$
Bayesian	9	46.8	47.2	0.8	0.997
Lease squares fitting	9	46.8	49.8	6.5	0.971

6.5 Comparison of the Models

The results of the tool wear, mean tool life values, and regression quality using exponential and Gompertz models are compared with the measured tool wear; see Table 6.4. According to the table, the regression quality and tool life prediction at $V_B = 0.3$ *mm*, are improved using sequential prediction method from 0 to 9 *deg* rake angles tools for both functions. Although both functions could predict the tool wear curve with the fit quality higher than $R^2 \sim 0.95$, the posterior exponential functions are more accurate than Gompertz function. The results of predicted values of the tool life using the posterior mean function and the corresponding error percentage at $V_B = 0.3$ *mm* are shown in Table 6.4. As demonstrated, Gompertz function could estimate the mean values of the tool life using 0 and 3 *deg* rake angles tools more accurate than the exponential function. Nevertheless, Exponential function estimate the mean tool life values using 9 *deg* rake angle tool more precise than Gompertz function.

Table 6.4: Comparison of the wear prediction accuracy using exponential and Gompertz functions

No.	*Function*	*Tool rake angle (deg)*	$T_{\_measured}$ *(min)*	$T_{\_predicted}$ *(min)*	$T_{\_error}$ *(%)*	$R_{\_Squared}$
1	Exponential	0	83.2	80.6	3	0.987
2	Exponential	3	63.8	65	1.8	0.991
3	Exponential	9	46.8	46.9	0.2	0.997
4	Gompertz	0	83.2	81.9	1.5	0.984
5	Gompertz	3	63.8	64.4	1	0.988
6	Gompertz	9	46.8	47.2	0.8	0.997

6.6 Conclusions

In this chapter, the sequential prediction of tool wear growth was performed using the Metropolis algorithm of the Bayesian method for the exponential and Gompertz models. MCMC simulations were implemented to identify the exponential and Gompertz model parameters and the wear curve prediction for the 0, 3, and 9 *deg* rake angle tools. First, the measured tool wear data of the steady-state and failure region were used for training of the model parameter and building of the probabilistic model using the 0 *deg* rake angle tool. Second, posteriors of the previous geometry were used for the new geometries (3 and 9 *deg* rake angles tools). In this scenario, the wear data of the steady-state region were only used for training of the model parameter, and then the probabilistic models predicted the tool wear data of the steady-state and failure regions. The probabilistic prediction result using the 9 *deg* rake angle tool was compared with the least squares fitting prediction. It was concluded that:

1. MCMC applications to both functions were completed successfully so that the prediction of tool wear, tool life, and the regression quality values were improved from 0 to 9 *deg* rake angles tools,
2. The Bayesian method can predict the wear curve more accurate (smaller error percentage) with the higher fit quality compared to the least squares fitting method. This is because of the integration of prior knowledge about the parameters to the analysis.

3. Bayesian Gompertz model could predict the tool life at $V_B = 0.3$ *mm*, more accurate than the exponential model for the first two geometries (0 and 3 *deg* rake angle tools).
4. The Bayesian exponential model could predict the tool life at $V_B = 0.3$ *mm,* slightly more accurate than the Gompertz model for the last geometries (9 *deg* rake angle tools).

Therefore, the Bayesian inference could identify the model parameters of the tool wear curves accurately, and the results could be used as an initial belief for the subsequent studies. Furthermore, the the Bayesian method requires fewer input values in the presence of informative priors and could evaluate the inherent uncertainties of the design and manufacturing process to the models.

7 Conclusions and Future Scope

The research in this dissertation focuses on the application of Bayesian inference as a predictive modeling approach to predict the machining and cutting tools performance metrics. Bayesian inference is a probabilistic modeling approach, which takes into account the inherent uncertainties due to the machining process, physical models, and measurement variabilities. The approach can incorporate initial belief or prior knowledge (*e.g.*, expert opinions, and previous experiments results) into the current and future analysis. In this way, the machining variables such as cutting forces, tool life and tool wear growth can be updated and used from one geometry of the cutting tools for another geometry in a sequential manner. The application of Bayesian inference to predict the performance metrics was studied in four chapters.

In chapter 3, the Bayesian inference was applied to the Merchant and Kienzle force models to predict the tangential and feed forces for two different tool geometries. It was shown that Kienzle model could predict the cutting forces using low feed values accurately, while the Merchant model is not. The results of the forces predictions were verified with the orthogonal turning data.

In chapter 4, Bayesian MCMC was applied to the extended Kienzle force model to isolate ploughing from cutting force in the orthogonal turning process. The model parameters were identified comparing Bayesian and Least Squares Curve Fitting (LSF) methods. It was demonstrated that the Bayesian approach could isolate and predict the ploughing force from cutting force with minimum input training data and inherent uncertainty, while the LSF method cannot predict the ploughing force component. Once again, the probabilistic model was verified with the results of the orthogonal turning process obtained under other cutting condition.

In chapter 5, tool life prediction and reliability analysis of cutting tools were performed applying Bayesian inference to the Taylor tool life model. It was demonstrated that the Bayesian MCMC approach could predict the tool life of two different tool geometries, sequentially. The probabilistic models were validated with the results of the milling tool life data other than training data.

The reliability analysis results were performed by reliability (quantitative) and hazard (qualitative) functions. The functions were used to analyze the effect of cutting forces and tool geometry on the cutting tool reliability.

In chapter 6, tool wear growth in milling process was predicted using the Bayesian MCMC method. Three types of tool geometries were considered for probabilistic sequential prediction process. Two types of degradation models, exponential and Gompertz functions, were taken into account to predict the tool wear curve. First, the model parameters of the exponential and Gompertz functions were trained using the wear data of the steady-state wear region. Next, the tool wear curve of the both steady-state and failure regions was predicted, successfully. The wear curve posterior functions were shown by the mean function and standard deviation credible intervals. The posterior functions were in good agreement with the measured tool wear error bars. According to the prediction results, both probabilistic functions can predict the wear growth with high fit quality. However, Gompertz function can predict the tool life at $V_B = 0.3\ mm$, more accurate than an exponential function. Finally, Tool wear growth prediction using Bayesian MCMC was compared to deterministic LSF method. The probabilistic method can predict the tool wear curve with the higher fit quality and more accurate than LSF method. This is because of the possibility of combining the prior knowledge to the analysis using the Bayesian method.

7.1 Future Scope: Digital Twin Technology

With the advent of commercial design packages, the product design process is being more and more digitalized. These programs enable designers to develop feature-based objects and simulate the behavior of the product. The digital programs (*e.g.*, CAD, CAM, CAE, FEA) reflect the virtual world of a product. In parallel to the virtual world, Internet of Things (IoT) platform enables the users to collect data from sensorized physical devises and transmit the data to the "cloud" and analyze those using data analytic programs. The IoT is heavily concentrated on the physical world. Given this background, the interaction between the virtual and physical worlds can be performed using digital twin, which is a new emerging and fast-growing technology. Traditionally, the virtual and physical products and spaces are built, analyzed, and verified separately from each other. However, in the digital twin framework, the virtual and physical spaces are not isolated, but there are two-way interacting channels, providing data exchange between the spaces [120].

The concept of the digital twin defined by Grieves at one of his presentations slides at the University of Michigan. The slide, as shown in Figure 7.1, was merely called "Conceptual Ideal for PLM." Nevertheless, it holds all the elements of the digital twin: real space, virtual space, the link for information flow from real space to virtual space and vice versa and virtual sub-spaces. The premise driving the model was that each system composed of two systems, the physical system that always exists and a new virtual model that holds all of the information about the physical system. This meant that there was a mirroring or twinning of systems between what existed in real space to what existed in virtual space, reciprocally [121].

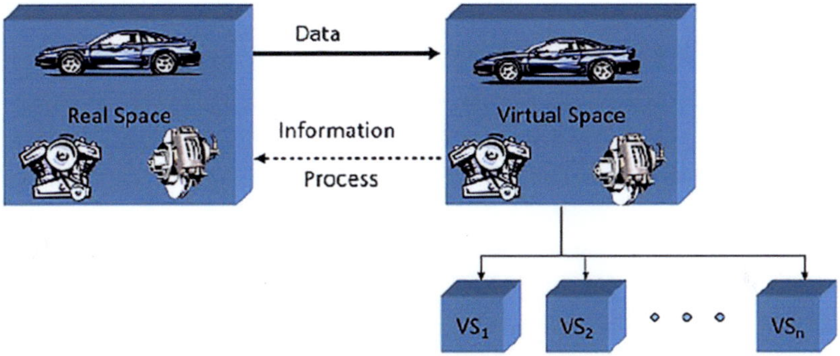

Figure 7.1: Conceptual ideal for PLM [121]

Since the introduction of this model, several explanations and definitions have been proposed [122]. Tuegel *et al.* [123] have used the digital twin for re-engineering of the aircraft structure and its life prediction. According to their explanation, digital twin can update data in real time, so that virtual models can experience continuous improvement by comparing virtual and physical spaces, parallelly. A general definition of the digital twin which has been recognized and broadly used so far has been presented by Glaessegen and Stargel [124] as follows:

"Digital Twin is integrated multi-physics modeling, probabilistic simulation of a complex product and uses the best available physical models, sensor updates, to mirror the life of its identical twin. Meanwhile, digital twin consists of three parts: physical product, virtual product, and linkage between the physical and virtual product. It is used as a bridge between the physical space and virtual space".

A digital twin is an incorporated model of an as-built product including physics, fatigue, lifecycle, sensor information, performance simulations, among others. It is intended to reflect all manufacturing defects and be continually updated to include wear-and-tear sustained while in use [125]. Digital twin allows companies or end users to have a complete digital footprint of the product, from design and development phase till the end of the product life. Therefore, digital twin has increasing attention by both industry and academia. The applications of digital twin in industrial and academic projects are summarized in the next section.

7.1.1 Applications of Digital Twin

Typical applications of digital twin to products health management (PHM) in the aerospace industry are reported in [120]. Seshadri *et al.* [126] proposed a damage characterization method based on a digital twin for aircraft structural health management, which demonstrated great advancement in predicting the damage location, size, and orientation. Gockel *et al.* [127] proposed Airframe Digital Twin (ADT) to assess the flight state which helps find the subsequent damage in a real-time way. Another Industrial form of a *hardware* twin is the *Iron Bird*, a ground-based engineering tool used in aircraft industries to incorporate, optimize and validate vital aircraft systems (Airbus Industries). Due to the increasing power of simulation technologies more and more physical components are replaced by virtual models in the Iron Bird. This allows for using the concept of an Iron Bird in earlier development cycles, even when some physical components are not yet available. Extending this idea along all phases of the lifecycle causes a complete digital model of the physical system [128]. Besides aircraft, General Electric pays attention to using digital twin to forecast product health in the product lifecycle, which can make operations and maintenance more accurate.

7.1.2 Digital Twin for Cutting Tool Lifecycle

Cutting tools are the most flexible elements and the basis of machining industry. The work dealing with machining processes such as cutting, drilling, turning, and milling require proper selection of cutting tools. Although modern machine tools enable the manufacturers to achieve higher feed rates and cutting speeds, market demand for machining of new materials such as Nickel-based superalloys, Titanium and hard to cut materials requires harder, tougher and more reliable cutting tools. The decision of the most appropriate cutting tool design and development process can deliver savings of as much as 15% on overall costs and improve machining productivity by 20%. Proper design and selection of cutting tools can also minimize downtime of production lines (*i.e.*, the time taken to replace worn tools) [129]. Researchers reported that 8% of total manufacturing costs are due to the cutting tools in the manufacturing phase of the product lifecycle. Therefore, industrial companies should also concentrate on improving the tool life cycle.

The complexity and variety of today's cutting tools are continually growing due to the industrial companies request and desire including cost reduction and innovations in technology, etc. In this context, conventional tool design, delivery and distribution, and management systems are not able to pursue a comprehensive approach to collect all necessary data along the tool life cycle. To provide solutions to these volatile conditions, processes within the cutting tool life cycle need to be monitored, adjusted and optimized comprehensively from the design phase until the recycling phase [130].

Traditionally, there is no consideration of an automated concept to extract tools performance data during their operational life. This means that, once the tools are delivered to the end user, they do not communicate back to their producers about their performance. Therefore, it is necessary to provide a platform to integrate the tools performance data into the virtual tools' models.

Digital twin can provide this platform by leveraging smart cutting tools and sensor network of the machine tools. In this context, new generation of intelligent machine tools, versatile and affordable sensors, and the smart cutting tools can help to monitor the tool performance throughout its service and operation life. The recorded data of the cutting tool performance can be fed back to the virtual tool model to have a footprint of the entire tool lifecycle. The digital twin of cutting tools can provide a near-real-time linkage between the physical and digital status of the tools. So, the tool failures can be visualized on the virtual copy in the digital twin concept. Using this technology, the tool designer can realize the features, and functionality of the cutting tools during their operational life. This helps the tool manufacturer to understand the customer's requirement and optimize the functionality of the tool more efficiently than the traditional approach.

Digital twin technology can provide flexibility, agility and lower cost of production for the cutting tool manufacturers. However, significant challenges are also encountered in the generation and collection of data from the floor. Automated data acquisition is critical for the implementation of digital twin, particularly, for small to midsize companies, which still rely on manual methods for data gathering. What is the first step and how this will be getting started? It can be a challenging task to get there, but the journey starts with a single step [2,131].

References

[1] Christoph Roser, Industry 4.0 – What works, what doesn't | AllAboutLean.com, (n.d.). https://www.allaboutlean.com/industry-4-0-potentials/.

[2] H. Ahuett-Garza, T. Kurfess, A brief discussion on the trends of habilitating technologies for Industry 4.0 and Smart manufacturing, Manuf. Lett. 15 (2018) 60–63. doi:10.1016/J.MFGLET.2018.02.011.

[3] L. Monostori, B. Kádár, T. Bauernhansl, S. Kondoh, S. Kumara, G. Reinhart, O. Sauer, G. Schuh, W. Sihn, K. Ueda, Cyber-physical systems in manufacturing, CIRP Ann. 65 (2016) 621–641. doi:10.1016/J.CIRP.2016.06.005.

[4] S. Jeschke, C. Brecher, H. Song, D.B. Rawat, Industrial Internet of Things and Cyber Manufacturing Systems, Springer International Publishing, 2017. doi:10.1007/978-3-319-42559-7.

[5] R. Gao, L. Wang, R. Teti, D. Dornfeld, S. Kumara, M. Mori, M. Helu, Cloud-enabled prognosis for manufacturing, CIRP Ann. 64 (2015) 749–772. doi:10.1016/J.CIRP.2015.05.011.

[6] M. Salehi, Indirect estimation of cutting force in milling machine tools, Italy, 2013. https://www.politesi.polimi.it/handle/10589/86361.

[7] M. Salehi, P. Albertelli, M. Goletti, F. Ripamonti, G. Tomasini, M. Monno, Indirect Model Based Estimation of Cutting Force and Tool Tip Vibrational Behavior in Milling Machines by Sensor Fusion, Procedia CIRP. 33 (2015) 239–244. doi:10.1016/J.PROCIR.2015.06.043.

[8] J. Karandikar, T. McLeay, S. Turner, T. Schmitz, Remaining Useful Tool Life Predictions Using Bayesian Inference, in: Vol. 2 Syst. Micro Nano Technol. Sustain. Manuf., ASME, 2013: p. V002T02A027. doi:10.1115/MSEC2013-1152.

[9] Y. Koren, X. Gu, W. Guo, Reconfigurable manufacturing systems: Principles, design, and future trends, Front. Mech. Eng. 13 (2018) 121-136. doi:10.1007/s11465-018-0483-0.

[10] M. Bortolini, E. Ferrari, M. Gamberi, F. Pilati, M. Faccio, Assembly system design in the Industry 4.0 era: a general framework, IFAC-PapersOnLine. 50 (2017) 5700–5705. doi:10.1016/J.IFACOL.2017.08.1121.

[11] T. Bauernhansl, T. Bauernhansl, Industry 4.0: Challenges and opportunities for the automation industry, in: 7th EFAC Assem. Technol. Conf., 2013: p. 23 Folien. http://publica.fraunhofer.de/documents/N-286876.html.

[12] M. Broy, Cyber-Physical Systems — Wissenschaftliche Herausforderungen Bei Der Entwicklung, in: Cyber-Physical Syst., Springer Berlin Heidelberg, Berlin, Heidelberg, 2010: pp. 17–31. doi:10.1007/978-3-642-14901-6_2.

[13] Product lifecycle management in industry 4.0 – Digital Factory – HANNOVER MESSE, Hann. Messe. (2016). http://www.hannovermesse.de/en/news/product-lifecycle-management-in-industry-4.0.xhtml.

[14] D. Bergsjö, Product Lifecycle Management-Architectural and Organisational Perspectives, n.d. www.chalmers.se (accessed August 10, 2018).

[15] C. Vila, J.V. Abellán-Nebot, J.C. Albiñana, G. Hernández, An Approach to Sustainable Product Lifecycle Management (Green PLM), Procedia Eng. 132 (2015) 585–592. doi:10.1016/J.PROENG.2015.12.608.

[16] Y. Altintas, P. Kersting, D. Biermann, E. Budak, B. Denkena, I. Lazoglu, Virtual process systems for part machining operations, CIRP Ann. 63 (2014) 585–605. doi:10.1016/J.CIRP.2014.05.007.

[17] A. Abdul Kadir, X. Xu, E. Hämmerle, Virtual machine tools and virtual machining—A technological review, Robot. Comput. Integr. Manuf. 27 (2011) 494–508. doi:10.1016/J.RCIM.2010.10.003.

[18] J.H. Ko, W.-S. Yun, D.-W. Cho, K.F. Ehmann, Development of a virtual machining system, part 1: approximation of the size effect for cutting force prediction, Int. J. Mach. Tools Manuf. 42 (2002) 1595–1605. doi:10.1016/S0890-6955(02)00137-2.

[19] W.-S. Yun, J.H. Ko, D.-W. Cho, K.F. Ehmann, Development of a virtual machining system, part 2: prediction and analysis of a machined surface error, Int. J. Mach. Tools Manuf. 42 (2002) 1607–1615. doi:10.1016/S0890-6955(02)00138-4.

[20] W.-S. Yun, J.H. Ko, H.U. Lee, D.-W. Cho, K.F. Ehmann, Development of a virtual machining system, part 3: cutting process simulation in transient cuts, Int. J. Mach. Tools Manuf. 42 (2002) 1617–1626. doi:10.1016/S0890-6955(02)00139-6.

[21] Y. Altintas, C. Brecher, M. Weck, S. Witt, Virtual Machine Tool, CIRP Ann. 54 (2005) 115–138. doi:10.1016/S0007-8506(07)60022-5.

[22] J.R. Evans, Business analytics : methods, models, and decisions, 1st ed., Pearson, Boston, 2013.

[23] Diagnostic Analytics, Cornerstone Glossary, (n.d.). https://www.cornerstoneondemand.com/glossary/diagnostic-analytics.

[24] I. Builders, Predictive Analytics for Manufacturing WebFOCUS RStat Optimizes Production and Process Efficiency, 2018. https://www.informationbuilders.com/sites/default/files/2018-01/fs_predictive_analytics_manufacturing_rstat_0.pdf.

[25] L. Calderone, Using Predictive Analytics in Manufacturing, Manufacturing Tomorrow, (n.d.). https://www.manufacturingtomorrow.com/article/2017/11/using-predictive-analytics-in-manufacturing/10544/.

[26] S. Finlay, Predictive analytics, data mining and big data : myths, misconceptions and methods, Palgrave Macmillan, Basingstoke, 2014.

[27] H. Ohlsson, J. Rydell, A. Brun, J. Roll, M. Andersson, A. Ynnerman, H. Knutsson, Enabling Bio-Feedback Using Real-Time fMRI, in: Proceedings of IEEE conference on decision and control, 2008: p. 3336. http://liu.diva-portal.org/smash/get/diva2:265503/FULLTEXT01.pdf.

[28] P.D. Hoff, A First Course in Bayesian Statistical Methods, Springer New York, New York, NY, 2009. doi:10.1007/978-0-387-92407-6.

[29] A.L. Samuel, Some Studies in Machine Learning Using the Game of Checkers, IBM J. Res. Dev. 3 (1959) 210–229. doi:10.1147/rd.33.0210.

[30] M. Cord, P. Cunningham, Machine learning techniques for multimedia : case studies on organization and retrieval, Springer, 2008.

[31] M. Awad, R. Khanna, Machine Learning, in: Effic. Learn. Mach., Apress, Berkeley, CA, 2015: pp. 1–18. doi:10.1007/978-1-4302-5990-9_1.

[32] I. Zaman, K. Pazouki, R. Norman, S. Younessi, S. Coleman, Challenges and Opportunities of Big Data Analytics for Upcoming Regulations and Future Transformation of the Shipping Industry, Procedia Eng. 194 (2017) 537–544. doi:10.1016/J.PROENG.2017.08.182.

[33] G.E.P. Box, Science and Statistics, J. Am. Stat. Assoc. 71 (1976) 791-799. http://www-sop.inria.fr/members/Ian.Jermyn/philosophy/writings/Boxonmaths.pdf.

[34] D.B. Dunson, Commentary: Practical Advantages of Bayesian Analysis of Epidemiologic Data, Am. J. Epidemiol. 153 (2001) 1222–1226. doi:10.1093/aje/153.12.1222.

[35] I.B. Wall, Probabilistic risk assessment in nuclear power plant regulation, Nucl. Eng. Des. 60 (1980) 11–24. doi:10.1016/0029-5493(80)90248-4.

[36] D. Hu, R. Wang, Z. Tao, Probabilistic design for turbine disk at high temperature, Aircr. Eng. Aerosp. Technol. 83 (2011) 199–207. doi:10.1108/00022661111138602.

[37] J. Sørensen, ; Dalsgaard, H.S. Toft, J.D. Sørensen, H.S. Toft, Probabilistic Design of Wind Turbines, Citation. 3 (2018) 241–257. doi:10.3390/en3020241.

[38] Y. Altintas, A. Ber, Manufacturing Automation: Metal Cutting Mechanics, Machine Tool Vibrations, and CNC Design, Cambridge University Press, 2001. doi:10.1115/1.1399383.

[39] K. Salonitis, A. Kolios, Reliability Assessment of Cutting Tools Life based on Advanced Approximation Methods, Procedia CIRP. 8 (2013) 397–402. doi:10.1016/J.PROCIR.2013.06.123.

[40] H. Wiklund, Bayesian and regression approaches to on-line prediction of residual tool life, Qual. Reliab. Eng. Int. 14 (1998) 303–309. doi:10.1002/(SICI)1099-1638(199809/10)14:5<303::AID-QRE155>3.0.CO;2-F.

[41] S. Kurada, C. Bradley, A review of machine vision sensors for tool condition monitoring, Comput. Ind. 34 (1997) 55–72. doi:10.1016/S0166-3615(96)00075-9.

[42] K. Salonitis, A. Kolios, Reliability assessment of cutting tool life based on surrogate approximation methods, Int. J. Adv. Manuf. Technol. 71 (2014) 1197–1208. doi:10.1007/s00170-013-5560-2.

[43] P.J. Arrazola, T. Özel, D. Umbrello, M. Davies, I.S. Jawahir, Recent advances in modelling of metal machining processes, CIRP Ann. 62 (2013) 695–718. doi:10.1016/J.CIRP.2013.05.006.

[44] M. Chandrasekaran, M. Muralidhar, C.M. Krishna, U.S. Dixit, Application of soft computing techniques in machining performance prediction and optimization: a literature review, Int. J. Adv. Manuf. Technol. 46 (2010) 445–464. doi:10.1007/s00170-009-2104-x.

[45] R.G. Khanchustambham, G.M. Zhang, A Neural Network Approach to On-line Monitoring of a Turning Process, (1992). https://drum.lib.umd.edu/handle/1903/5206 (accessed August 10, 2018).

[46] B.Y. Lee, Y.S. Tarng, S.C. Ma, Modeling of the process damping force in chatter vibration, Int. J. Mach. Tools Manuf. 35 (1995) 951–962. doi:10.1016/0890-6955(94)00046-M.

[47] R.. Dutta, S. Paul, A.. Chattopadhyay, Applicability of the modified back-propagation algorithm in tool condition monitoring for faster convergence, J. Mater. Process. Technol. 98 (2000) 299–309. doi:10.1016/S0924-0136(99)00295-2.

[48] R. Quiza, L. Figueira, J. Paulo Davim, Comparing statistical models and artificial neural networks on predicting the tool wear in hard machining D2 AISI steel, Int. J. Adv. Manuf. Technol. 37 (2008) 641-648. doi:10.1007/s00170-007-0999-7.

[49] B. Denkena, D. Biermann, Cutting edge geometries, CIRP Ann. – Manuf. Technol. 63 (2014) 631–653. doi:10.1016/j.cirp.2014.05.009.

[50] W.J. Endres, R.K. Kountanya, The effects of corner radius and edge radius on tool flank wear, J. Manuf. Process. 4 (2002) 89–96. doi:10.1016/S1526-6125(02)70135-7.

[51] E. Bassett, J. Köhler, B. Denkena, On the honed cutting edge and its side effects during orthogonal turning operations of AISI1045 with coated WC-Co inserts, CIRP J. Manuf. Sci. Technol. 5 (2012) 108–126. doi:10.1016/j.cirpj.2012.03.004.

[52] B. Denkena, J. Köhler, B. Breidenstein, A.M. Abrão, C.E.H. Ventura, Influence of the cutting edge preparation method on characteristics and performance of PVD coated carbide inserts in hard turning, Surf. Coatings Technol. 254 (2014) 447–454. doi:10.1016/j.surfcoat.2014.07.003.

[53] C.F. Wyen, K. Wegener, Influence of cutting edge radius on cutting forces in machining titanium, CIRP Ann. - Manuf. Technol. 59 (2010) 93–96. doi:10.1016/j.cirp.2010.03.056.

[54] K.D. Bouzakis, E. Bouzakis, S. Kombogiannis, S. Makrimallakis, G. Skordaris, N. Michailidis, P. Charalampous, R. Paraskevopoulou, R. M'Saoubi, J.C. Aurich, F. Barthelmä, D. Biermann, B. Denkena, D. Dimitrov, S. Engin, B. Karpuschewski, F. Klocke, T. Özel, G. Poulachon, J. Rech, V. Schulze, L. Settineri, A. Srivastava, K. Wegener, E. Uhlmann, P. Zeman, Effect of cutting edge preparation of coated tools on their performance in milling various materials, CIRP J. Manuf. Sci. Technol. 7 (2014) 264–273. doi:10.1016/j.cirpj.2014.05.003.

[55] R. van de Schoot, D. Kaplan, J. Denissen, J.B. Asendorpf, F.J. Neyer, M.A.G. van Aken, A gentle introduction to bayesian analysis: applications to developmental research., Child Dev. 85 (2014) 842–860. doi:10.1111/cdev.12169.

[56] F.A. Niaki, A Probabilistic-Based Approach to Monitoring Tool Wear State and Assessing Its Effect on Workpiece Quality in Nickel-Based Alloys, Clemson University, 2016. http://tigerprints.clemson.edu/all_dissertations/1858.

[57] A. Gelman, Prior Distribution, Encycl. Environmetrics. 3 (2006) 1634-1637. doi:10.1002/9780470057339.vap039.

[58] J. V. Beck, K.J. Arnold, Parameter estimation in engineering and science, John Wiley & Sons, 1977. https://www.thermalfluidscentral.org/e-books/book-intro.php?b=43.

[59] A. Eshky, Bayesian Methods of Parameter Estimation, n.d. http://homepages.inf.ed.ac.uk/rbf/CVonline/LOCAL_COPIES/AV0809/eshky.pdf.

[60] M. Steyvers, M. Chain, M. Carlo, Computational Statistics with Matlab, (2010).

[61] P. Boedeker, Hierarchical Linear Modeling with Maximum Likelihood, Restricted Maximum Likelihood, and Fully Bayesian Estimation. Practical Assessment, Research & Evaluation, 22 (2017). http://pareonline.net/getvn.asp?v=22&n=2.

[62] G. Sparacino, C. Tombolato, C. Cobelli, Maximum-likelihood versus maximum a posteriori parameter estimation of physiological system models: The c-peptide impulse response case study, IEEE Trans. Biomed. Eng. 47 (2000) 801–811. doi:10.1109/10.844232.

[63] C. Andrieu, N. De Freitas, A. Doucet, M.I. Jordan, An introduction to MCMC for machine learning, Mach. Learn. 50 (2003) 5–43. doi:10.1023/A:1020281327116.

[64] H. Fischer, The Central Limit Theorem from Laplace to Cauchy: Changes in Stochastic Objectives and in Analytical Methods, in: A Hist. Cent. Limit Theorem. Sources Stud. Hist. Math. Phys. Sci., Springer New York, New York, NY, 2010: pp. 1–29. doi:10.1007/978-0-387-87857-7.

[65] J.M. Karandikar, the Fundamental Application of Decision Analysis To Manufacturing, University of North Carolina at Charlotte, 2013. http://libres.uncg.edu/ir/uncc/f/Karandikar_uncc_0694D_10487.pdf.

[66] G.O. Roberts, J.S. Rosenthal, Optimal scaling for various Metropolis-Hastings algorithms, Stat. Sci. 16 (2001) 351–367. doi:10.1214/ss/1015346320.

[67] F. Akhavan Niaki, D. Ulutan, L. Mears, Parameter Inference Under Uncertainty in End-Milling γ'-Strengthened Difficult-to-Machine Alloy, J. Manuf. Sci. Eng. 138 (2016) 061014. doi:10.1115/1.4033041.

[68] J. Geweke, Evaluating the accuracy of sampling-based approaches to the calculation of posterior moments, Bayesian Stat. 4. 4 (1992) 169-193. doi:1176289.

[69] J.M. Karandikar, A.E. Abbas, T.L. Schmitz, Tool life prediction using Bayesian updating. Part 2: Turning tool life using a Markov Chain Monte Carlo approach, Precis. Eng. 38 (2014) 18–27. doi:10.1016/J.PRECISIONENG.2013.06.007.

[70] M.E. Merchant, Mechanics of the metal cutting process. II. Plasticity conditions in orthogonal cutting, J. Appl. Phys. 16 (1945) 318–324. doi:10.1063/1.1707596.

[71] E. Shamoto, Y. Altıntas, Prediction of Shear Angle in Oblique Cutting with Maximum Shear Stress and Minimum Energy Principles, J. Manuf. Sci. Eng. 121 (1999) 399. doi:10.1115/1.2832695.

[72] D.W. Smithey, S.G. Kapoor, R.E. DeVor, A NEW MECHANISTIC MODEL FOR PREDICTING WORN TOOL CUTTING FORCES, Mach. Sci. Technol. 5 (2001) 23–42. doi:10.1081/MST-100103176.

[73] T.L. Schmitz, K.S. Smith, M. Dynamics, Milling Dynamics, in: Milling, Springer US, Boston, MA, 2009. doi:10.1007/978-0-387-09645-2.

[74] T.L. Schmitz, J. Karandikar, N. Ho Kim, A. Abbas, Uncertainty in Machining: Workshop Summary and Contributions, J. Manuf. Sci. Eng. 133 (2011) 051009. doi:10.1115/1.4004923.

[75] P. Mehta, M. Kuttolamadom, L. Mears, Mechanistic force model for machining process—theory and application of Bayesian inference, Int. J. Adv. Manuf. Technol. 91 (2017) 3673–3682. doi:10.1007/s00170-017-0064-0.

[76] E. Gözü, Y. Karpat, Uncertainty analysis of force coefficients during micromilling of titanium alloy, Int. J. Adv. Manuf. Technol. 93 (2017) 839–855. doi:10.1007/s00170-017-0567-8.

[77] M. Weber, T. Hochrainer, P. Gumbsch, H. Autenrieth, L. Delonnoy, V. Schulze, D. Löhe, J. Kotschenreuther, J. Fleischer, Investigation of size-effects in machining with geometrically defined cutting edges, Mach. Sci. Technol. 11 (2007) 447–473. doi:10.1080/10910340701697086.

[78] F. Vollertsen, D. Biermann, H.N. Hansen, I.S. Jawahir, K. Kuzman, Size effects in manufacturing of metallic components, CIRP Ann. - Manuf. Technol. 58 (2009) 566–587. doi:10.1016/j.cirp.2009.09.002.

[79] F. Klocke, O. Adams, T. Auerbach, S. Gierlings, S. Kamps, S. Rekers, D. Veselovac, M. Eckstein, A. Kirchheim, M. Blattner, R. Thiel, D. Kohler, New concepts of force measurement systems for specific machining processes in aeronautic industry, CIRP J. Manuf. Sci. Technol. 9 (2015) 31–38. doi:10.1016/J.CIRPJ.2015.01.006.

[80] R.W. Ivester, M. Kennedy, M. Davies, R. Stevenson, J. Thiele, R. Furness, S. Athavale, Assessment of machining models: progress report, Mach. Sci. Technol. 4 (2000) 511–538. doi:10.1080/10940340008945720.

[81] R.J. Schimmel, W.J. Endres, R. Stevenson, Application of an Internally Consistent Material Model to Determine the Effect of Tool Edge Geometry in Orthogonal Machining, J. Manuf. Sci. Eng. 124 (2002) 536. doi:10.1115/1.1448334.

[82] B. Denkena, H.K. Tönshoff, Spanen : Grundlagen, Springer-Verlag Berlin Heidelberg, 2011.

[83] M. Salehi, T.L. Schmitz, R. Copenhaver, R. Haas, J. Ovtcharova, Probabilistic Prediction of Cutting and Ploughing Forces using Extended Kienzle Force Model in Orthogonal Turning Process, Procedia CIRP. 77 (2018) 90–93. doi:10.1016/J.PROCIR.2018.08.228.

[84] D.J. Waldorf, R.E. DeVor, S.G. Kapoor, An Evaluation of Ploughing Models for Orthogonal Machining, J. Manuf. Sci. Eng. 121 (1999) 550. doi:10.1115/1.2833050.

[85] P. Albrecht, New Developments in the Theory of the Metal-Cutting Process: Part I. The Ploughing Process in Metal Cutting, J. Eng. Ind. 82 (1960) 348. doi:10.1115/1.3664242.

[86] D.J. Waldorf, A simplified model for ploughing forces in turning, J. Manuf. Process. 8 (2006) 76–82. doi:10.1016/S1526-6125(07)00005-9.

[87] ISO 3685:1993 - Tool-life testing with single-point turning tools, (n.d.).

[88] V.P. Astakhov, J.P. Davim, Tools (Geometry and Material) and Tool Wear, in: Machining, Springer London, London, 2008: pp. 29–57. doi:10.1007/978-1-84800-213-5_2.

[89] Y.-C. Yen, J. Söhner, B. Lilly, T. Altan, Estimation of tool wear in orthogonal cutting using the finite element analysis, J. Mater. Process. Technol. 146 (2004) 82–91. doi:10.1016/S0924-0136(03)00847-1.

[90] V. Schulze, F. Zanger, Development of a Simulation Model to Investigate Tool Wear in Ti-6Al-4V Alloy Machining, Adv. Mater. Res. 223 (2011) 535–544. doi:10.4028/www.scientific.net/AMR.223.535.

[91] M. Calamaz, J. Limido, M. Nouari, C. Espinosa, D. Coupard, M. Salaün, F. Girot, R. Chieragatti, Toward a better understanding of tool wear effect through a comparison between experiments and SPH numerical modelling of machining hard materials, Int. J. Refract. Met. Hard Mater. 27 (2009) 595–604. doi:10.1016/J.IJRMHM.2008.09.005.

[92] A. Attanasio, D. Umbrello, Abrasive and diffusive tool wear FEM simulation, Int. J. Mater. Form. 2 (2009) 543–546. doi:10.1007/s12289-009-0475-z.

[93] I.S. Jawahir, R. Ghosh, A.K. Balaji, P.X. Li, Predictability of tool failure modes in turning with complex grooved tools using the equivalent toolface (ET) model, Wear. 244 (2000) 94–103. doi:10.1016/S0043-1648(00)00447-6.

[94] J.M. Karandikar, A.E. Abbas, T.L. Schmitz, Tool life prediction using Bayesian updating. Part 1: Milling tool life model using a discrete grid method, Precis. Eng. 38 (2014) 9–17. doi:10.1016/J.PRECISIONENG.2013.06.006.

[95] J.M. Karandikar, A.E. Abbas, T.L. Schmitz, Tool Life Prediction using Random Walk Bayesian Updating, Mach. Sci. Technol. 17 (2013) 410-442. doi:10.1080/10910344.2013.806103.

[96] F.A. Niaki, D. Ulutan, L. Mears, Parameter Estimation Using Markov Chain Monte Carlo Method in Mechanistic Modeling of Tool Wear During Milling, in: Vol. 2 Mater. Biomanufacturing; Prop. Appl. Syst. Sustain. Manuf., ASME, 2015: p. V002T04A006. doi:10.1115/MSEC2015-9357.

[97] T.I. El Wardany, M.A. Elbestawi, Prediction of tool failure rate in turning hardened steels, Int. J. Adv. Manuf. Technol. 13 (1997) 1–16. doi:10.1007/BF01179225.

[98] A. Jeang, Reliable tool replacement policy for quality and cost, Eur. J. Oper. Res. 108 (1998) 334–344. doi:10.1016/S0377-2217(96)00368-2.

[99] K.-S. Wang, W.-S. Lin, F.-S. Hsu, A New Approach for Determining the Reliability of a Cutting Tool, Int. J. Adv. Manuf. Technol. 17 (2001) 705–709. doi:10.1007/s001700170114.

[100] Z. Klim, E. Ennajimi, M. Balazinski, C. Fortin, Cutting tool reliability analysis for variable feed milling of 17-4PH stainless steel, Wear. 195 (1996) 206–213. doi:10.1016/0043-1648(95)06863-5.

[101] F. Ding, Z. He, Cutting tool wear monitoring for reliability analysis using proportional hazards model, Int. J. Adv. Manuf. Technol. 57 (2011) 565–574. doi:10.1007/s00170-011-3316-4.

[102] C.J. Cortés Rodríguez, Cutting edge preparation of precision cutting tools by applying micro-abrasive jet machining and brushing, Kassel Univ. Press, 2009. http://www.upress.uni-kassel.de/katalog/abstract_en.php?978-3-89958-712-8.

[103] D.M. Kennedy, J. Vahey, D. Hanney, Micro shot blasting of machine tools for improving surface finish and reducing cutting forces in manufacturing, Mater. Des. 26 (2005) 203–208. doi:10.1016/J.MATDES.2004.02.013.

[104] U.D. Kumar, J. Crocker, J. Knezevic, M. El-Haram, Reliability, Maintenance, and Logistic Support — Introduction, in: Reliab. Maint. Logist. Support, Springer US, Boston, MA, 2000: pp. 1–12. doi:10.1007/978-1-4615-4655-9_1.

[105] P.A. Dearnley, E.M. Trent, Wear mechanisms of coated carbide tools, Met. Technol. 9 (1982) 60–75. doi:10.1179/030716982803285909.

[106] K.C. Ee, P.X. Li, A.K. Balaji, I.S. Jawahir, R. Stevenson, Performance-Based Predictive Models and Optimization Methods for Turning Operations and Applications: Part 1—Tool Wear/Tool Life in Turning with Coated Grooved Tools, J. Manuf. Process. 8 (2006) 54–66. doi:10.1016/S1526-6125(06)70102-5.

[107] P. Mathew, Use of predicted cutting temperatures in determining tool performance, Int. J. Mach. Tools Manuf. 29 (1989) 481–497. doi:10.1016/0890-6955(89)90066-7.

[108] A. Molinari, M. Nouari, Modeling of tool wear by diffusion in metal cutting, Wear. 252 (2002) 135–149. doi:10.1016/S0043-1648(01)00858-4.

[109] E. Usui, T. Shirakashi, T. Kitagawa, Analytical prediction of cutting tool wear, Wear. 100 (1984) 129–151. doi:10.1016/0043-1648(84)90010-3.

[110] T. Matsumura, T. Shirakashi, E. Usui, Identification of Wear Characteristics in Tool Wear Model of Cutting Process, Int. J. Mater. Form. 1 (2008) 555–558. doi:10.1007/s12289-008-0297-4.

[111] H. Takeyama, R. Murata, Basic Investigation of Tool Wear, J. Eng. Ind. 85 (1963) 33. doi:10.1115/1.3667575.

[112] A. Attanasio, E. Ceretti, A. Fiorentino, C. Cappellini, C. Giardini, Investigation and FEM-based simulation of tool wear in turning operations with uncoated carbide tools, Wear. 269 (2010) 344–350. doi:10.1016/J.WEAR.2010.04.013.

[113] L. Filice, F. Micari, L. Settineri, D. Umbrello, Wear modelling in mild steel orthogonal cutting when using uncoated carbide tools, Wear. 262 (2007) 545–554. doi:10.1016/J.WEAR.2006.06.022.

[114] L.-J. Xie, J. Schmidt, C. Schmidt, F. Biesinger, 2D FEM estimate of tool wear in turning operation, Wear. 258 (2005) 1479–1490. doi:10.1016/J.WEAR.2004.11.004.

[115] A. Attanasio, E. Ceretti, S. Rizzuti, D. Umbrello, F. Micari, 3D finite element analysis of tool wear in machining, CIRP Ann. 57 (2008) 61-64. doi:10.1016/J.CIRP.2008.03.123.

[116] A. Attanasio, E. Ceretti, C. Giardini, L. Filice, D. Umbrello, Criterion to evaluate diffusive wear in 3D simulations when turning AISI 1045 steel, Int. J. Mater. Form. 1 (2008) 495–498. doi:10.1007/s12289-008-0130-0.

[117] T. Özel, Y. Karpat, Predictive modeling of surface roughness and tool wear in hard turning using regression and neural networks, Int. J. Mach. Tools Manuf. 45 (2005) 467–479. doi:10.1016/J.IJMACHTOOLS.2004.09.007.

[118] A. Attanasio, E. Ceretti, C. Giardini, Analytical Models for Tool Wear Prediction During AISI 1045 Turning Operations, Procedia CIRP. 8 (2013) 218–223. doi:10.1016/J.PROCIR.2013.06.092.

[119] F. Klocke, P. Frank, Simulation of tool wear in hard turning, in: 9th CIRP Int. Work. Model. Mach. Oper., 2006.

[120] F. Tao, F. Sui, A. Liu, Q. Qi, M. Zhang, B. Song, Z. Guo, S.C.-Y. Lu, A.Y.C. Nee, Digital twin-driven product design framework, Int. J. Prod. Res. (2018) 1–19. doi:10.1080/00207543.2018.1443229.

[121] M. Grieves, J. Vickers, Digital Twin: Mitigating Unpredictable, Undesirable Emergent Behavior in Complex Systems, in: Transdiscipl. Perspect. Complex Syst., Springer International Publishing, Cham, 2017: pp. 85–113. doi:10.1007/978-3-319-38756-7_4.

[122] F. Tao, J. Cheng, Q. Qi, M. Zhang, H. Zhang, F. Sui, Digital twin-driven product design, manufacturing and service with big data, Int. J. Adv. Manuf. Technol. 94 (2018) 3563–3576. doi:10.1007/s00170-017-0233-1.

[123] E.J. Tuegel, A.R. Ingraffea, T.G. Eason, S.M. Spottswood, Reengineering Aircraft Structural Life Prediction Using a Digital Twin, Int. J. Aerosp. Eng. 2011 (2011) 1–14. doi:10.1155/2011/154798.

[124] E.H. Glaessgen, D.S. Stargel, The Digital Twin Paradigm for Future NASA and U.S. Air Force Vehicles, in: 53rd Struct. Struct. Dyn. Mater. Conf. Spec. Sess. Digit. Twin, 2012. https://ntrs.nasa.gov/search.jsp?R=20120008178.

[125] J.B. Reid, D.H. Rhodes, Digital System Models: An investigation of the non-technical challenges and research needs, in: 2016 Conf. Syst. Eng. Res., 2016. http://seari.mit.edu/documents/preprints/REID_CSER16.pdf.

[126] B.R. Seshadri, T. Krishnamurthy, Structural Health Management of Damaged Aircraft Structures Using Digital Twin Concept, in: 25th AIAA/AHS Adapt. Struct. Conf., American Institute of Aeronautics and Astronautics, Reston, Virginia, 2017. doi:10.2514/6.2017-1675.

[127] B. Gockel, A. Tudor, M. Brandyberry, R. Penmetsa, E. Tuegel, Challenges with Structural Life Forecasting Using Realistic Mission Profiles, in: 53rd AIAA/ASME/ASCE/AHS/ASC Struct. Struct. Dyn. Mater. Conf. AIAA/ASME/AHS Adapt. Struct. Conf. AIAA, American Institute of Aeronautics and Astronautics, Reston, Virigina, 2012. doi:10.2514/6.2012-1813.

[128] R. Rosen, G. von Wichert, G. Lo, K.D. Bettenhausen, About The Importance of Autonomy and Digital Twins for the Future of Manufacturing, IFAC-PapersOnLine. 48 (2015) 567–572. doi:10.1016/J.IFACOL.2015.06.141.

[129] G. Drori, Proper Cutting Tool Choice is Vital to Productivity: Production Machining, Prod. Mach. (2015). https://www.productionmachining.com/columns/proper-cutting-tool-choice-is-vital-to-productivity.

[130] D. Brenner, F. Kleinert, J. Imiela, E. Westkämper, Life Cycle Management of Cutting Tools: Comprehensive Acquisition and Aggregation of Tool Life Data, Procedia CIRP. 61 (2017) 311–316. doi:10.1016/J.PROCIR.2016.11.168.

[131] A. Parrott, Industry 4.0 and the digital twin Manufacturing meets its match A Deloitte series on Industry 4.0, digital manufacturing enterprises, and digital supply networks, Deloitte, Univ. Press. (2017). https://www2.deloitte.com/content/dam/Deloitte/cn/Documents/cip/deloitte-cn-cip-industry-4-0-digital-twin-technology-en-171215.pdf.

REIHE INFORMATIONSMANAGEMENT IM ENGINEERING KARLSRUHE

Karlsruher Institut für Technologie | ISSN 1860-5990

Herausgeber:
Institut für Informationsmanagement im Ingenieurwesen (IMI)
o. Prof. Dr. Dr.-Ing. Dr. h.c. Jivka Ovtcharova

Band 1 – 2005 **SEIDEL, MICHAEL**
Methodische Produktplanung. Grundlagen, Systematik und Anwendung im Produktentstehungsprozess. 2005
ISBN 3-937300-51-1

Band 1 – 2006 **PRIEUR, MICHAEL**
Functional elements and engineering template-based product development process. Application for the support of stamping tool design. 2006
ISBN 3-86644-033-2

Band 2 – 2006 **GEIS, STEFAN RAFAEL**
Integrated methodology for production related risk management of vehicle electronics (IMPROVE). 2006
ISBN 3-86644-011-1

Band 1 – 2007 **GLOSSNER, MARKUS**
Integrierte Planungsmethodik für die Presswerkneutypplanung in der Automobilindustrie. 2007
ISBN 978-3-86644-179-8

Band 2 – 2007 **MAYER-BACHMANN, ROLAND**
Integratives Anforderungsmanagement. Konzept und Anforderungsmodell am Beispiel der Fahrzeugentwicklung. 2008
ISBN 978-3-86644-194-1

Band 1 – 2008 **MBANG SAMA, ACHILLE**
Holistic integration of product, process and resources integration in the automotive industry using the example of car body design and production. Product design, process modeling, IT implementation and potential benefits. 2008
ISBN 978-3-86644-243-6

Die Bände sind unter www.ksp.kit.edu als PDF frei verfügbar oder als Druckausgabe bestellbar.

Band 2 – 2008	**WEIGT, MARKUS** Systemtechnische Methodenentwicklung : Diskursive Definition heuristischer prozeduraler Prozessmodelle als Beitrag zur Bewältigung von informationeller Komplexität im Produktleben. 2008 ISBN 978-3-86644-285-6
Band 1 – 2009	**KRAPPE, HARDY** Erweiterte virtuelle Umgebungen zur interaktiven, immersiven Verwendung von Funktionsmodellen. 2009 ISBN 978-3-86644-380-8
Band 2 – 2009	**ROGALSKI, SVEN** Entwicklung einer Methodik zur Flexibilitätsbewertung von Produktionssystemen. Messung von Mengen-, Mix- und Erweiterungsflexibilität zur Bewältigung von Planungsunsicherheiten in der Produktion. 2009 ISBN 978-3-86644-383-9
Band 3 – 2009	**FORCHERT, THOMAS M.** Prüfplanung. Ein neues Prozessmanagement für Fahrzeugprüfungen. 2009 ISBN 978-3-86644-385-3
Band 1 – 2011	**ERKAYHAN, ŞEREF** Ein Vorgehensmodell zur automatischen Kopplung von Services am Beispiel der Integration von Standardsoftwaresystemen. 2011 ISBN 978-3-86644-697-7
Band 2 – 2011	**MEIER, GUNTER** Prozessintegration des Target Costings in der Fertigungsindustrie am Beispiel Sondermaschinenbau. 2011 ISBN 978-3-86644-679-3
Band 1 – 2012	Nicht erschienen
Band 2 – 2012	**WUTTKE, FABIAN** Robuste Auslegung von Mehrkörpersystemen. Frühzeitige Robustheitsoptimierung von Fahrzeugmodulen im Kontext modulbasierter Entwicklungsprozesse. 2012 ISBN 978-3-86644-896-4

Band 3 – 2012	**KATIČIĆ, JURICA** Methodik für Erfassung und Bewertung von emotionalem Kundenfeedback für variantenreiche virtuelle Produkte in immersiver Umgebung. 2012 ISBN 978-3-86644-930-5
Band 1 – 2013	**LOOS, MANUEL NORBERT** Daten- und termingesteuerte Entscheidungsmethodik der Fabrikplanung unter Berücksichtigung der Produktentstehung. 2013 ISBN 978-3-86644-963-3
Band 2 – 2013	**SYAL, GAGAN** CAE - PROCESS AND NETWORK: A methodology for continuous product validation process based on network of various digital simulation methods. 2013 ISBN 978-3-7315-0090-2
Band 1 – 2016	**BURGER, ALEXANDER** Design for Customer: Methodik für nachhaltige Kundenlösungen unter Zuhilfenahme eines bedürfnisorientierten Leistungskonfigurators. 2016 ISBN 978-3-7315-0168-8
Band 2 – 2016	**HOPF, JENS MICHAEL** Framework for the Integration of Mobile Device Features in PLM. 2016 ISBN 978-3-7315-0498-6
Band 1 – 2017	**WALLA, WALDEMAR** Standard- und Modulbasierte digitale Rohbauprozesskette: Frühzeitige Produktbeeinflussung bezüglich Produktionsanforderungen im Karosserierohbau der Automobilindustrie. 2017 ISBN 978-3-7315-0600-3
Band 1 – 2018	**WEISER, ANN-KATRIN** Methodik eines holistischen Variantenmanagements modularer Produktfamilien – Grundlagen, Systematik und beispielhafte Anwendung der VM_{ahead} Methodik. 2018 ISBN 978-3-7315-0775-8

Band 2 – 2018
STANEV, STILIAN
Methodik zur produktionsorientierten Produktanalyse für die Wiederverwendung von Produktionssystemen – 2REUSE. Konzept, Informationsmodell und Validierung am besonderen Beispiel des Karosserierohbaus in der Automobilindustrie. 2018
ISBN 978-3-86644-932-9

Band 1 – 2019
KLINGER, JULIUS FRIEDRICH
Tolerance Simulation in the Loop: Ansätze zur Verbesserung der Vorhersagegenauigkeit der Toleranzsimulation im Automobilbau durch Adaption an reale Fertigungsprozesse. 2019
ISBN 978-3-7315-0876-2

Band 2 – 2019
SALEHI, MEHDI
Bayesian-Based Predictive Analytics for Manufacturing Performance Metrics in the Era of Industry 4.0. 2019
ISBN 978-3-7315-0908-0